IMAGES
of Aviation

McChord Field

Welcome to McChord! This iconic c. 1951 photograph of the main gate of McChord Field shows Mount Rainier in the background. (Courtesy of McChord Museum.)

On the Cover: A crowd gathers around Brig. Gen. B.K. Yount, assistant chief of the Army Air Corps, during the dedication ceremony officially opening McChord Field on July 3, 1940, in conjunction with the Tacoma Narrows Bridge. At the time of the dedication, there were 1,821 enlisted men and 203 officers stationed at the base. Approximately 10,000 people attended the dedication ceremony, which included a flyover by Douglas B-18 Bolos and B-23 Dragon Bombers. The planes were also on static display for public viewing. When the field officially opened, there were about 70 medium twin-motored bombers, assigned to the base. (Courtesy of Tacoma Public Library.)

IMAGES
of Aviation

McCHORD FIELD

Kimberly Peterson

ARCADIA
PUBLISHING

ISBN 978-0-7385-9971-7

Published by Arcadia Publishing
Charleston, South Carolina

Printed in the United States of America

Library of Congress Control Number: 2012952523

For all general information, please contact Arcadia Publishing:
Telephone 843-853-2070
Fax 843-853-0044
E-mail sales@arcadiapublishing.com
For customer service and orders:
Toll-Free 1-888-313-2665

Visit us on the Internet at www.arcadiapublishing.com

Dedicated to all the airmen and their families of the past, present, and future who, while stationed at McChord Field, defend and protect our nation

Contents

ACKNOWLEDGMENTS

I have benefited from the generosity and willingness of many to help me in this journey to trace the contrails of McChord Field's history. When my family and I did a Permanent Change of Station (PCS) to McChord Field, my understanding of the base's role in history was limited. As these pages and the timeline have been completed and my understanding of McChord Field's history has grown, I would like to express my gratitude to the staff at the Air Force Historical Research Agency, Maxwell Air Force Base, Alabama. They were always willing to locate and bring to me boxes full of base histories, maps, and photographs, making my weeklong stay very enjoyable. Another source of photographs came from the Tacoma Public Library, Jodi Gripp and Jeanie Fisher, two wonderful staff people in the Northwest Reading Room, who greatly helped with locating files and photographs about the base. I would also like to thank Raymond Jordan and Ernest White at the McChord Museum. Ray's in-depth knowledge of McChord's aircraft and history is inspiring. Ernie's enthusiasm for all things McChord and airplanes is very catching, and I thank him for his willingness to help me on this project. To my children, who always get excited to see an airplane flying in the sky, your joy always brings a smile to my face. Lastly, to my husband, Pete, the bulk of this book was written while he was away, defending our nation. Even though we were apart, his support, words of wisdom, and love made the distance seem not so far, giving me the courage to complete this history.

INTRODUCTION

Nestled between the Cascade Mountains and Puget Sound in Pierce County, Washington, McChord Field is a part of Joint Base Lewis-McChord, formerly known as McChord Air Force Base. McChord Field, home to the 62nd Airlift Wing, is located in Tacoma, 45 miles south of Seattle, and has a very rich and detailed history at home and around the world.

McChord Field was not always full of concrete, enormous hangars, and gigantic aircraft. Not so long ago, fields covered with trees and grass occupied the site. Ownership of the land had passed from the Northwest Fur Company to the Hudson's Bay Company in 1821. In 1833, the Hudson Bay Company founded Fort Nisqually and began to operate a number of farms and sheep stations on 150,000 acres of land. Around 1847, the company established a sheep station in the area where the east gate is now located. Nearly 30 years later, the company sold its lands to the US government, ending the operation of the sheep stations.

Early in the 20th century, the sheep fields were replaced by an airfield. On November 7, 1927, a bond issued by Pierce County provided the funds to establish an airfield, which was voted and approved by 66 percent of the electorate. After much debate, the Pierce County Commissioners voted to name the new airfield Tacoma Field. The geography of Tacoma Field and the surrounding area was swamp, plains, and prairie, but the location, weather, and terrain still offered a good location to build the airfield. The temperature is moderate year-round, the land is flat, ranging from 200 to 800 feet above sea level, and the Cascade and Olympic Mountain ranges provide shelter from the northwesterly winds.

Despite its good location, Tacoma Field was operating in the red in early 1933. The Army became interested in Tacoma Field and sent 100 Army airships from San Francisco in May to take part in a Fort Lewis training exercise. This move helped save the airport, but around 1934, Congress authorized the secretary of war to establish airfields in six strategic areas around the United States. The 900-acre Tacoma Field was chosen as the best location in the Pacific Northwest.

This deal was a preview of the fate of the airfield. On May 5, 1938, the governor of Washington approved the transfer of land to the US government. The new owner of the old Tacoma airport now owned one hangar and two landing strips. Tacoma Field received a new name, McChord Field, in honor of Col. William C. McChord, who had been killed in an airplane accident near Richmond, Virginia, on August 18, 1937. Shortly afterward, contracts were awarded and construction crews began to shape and mold many of the structures that remain in place to this very day. Hangars No. 1, 2, 3, and 4 were built, as were the enormous Air Corps barracks, also known as the Castle. Also erected were warehouses, housing for enlisted and officer personnel, and everything necessary to operate a base.

At last, McChord Field was ready for the public to view all of its latest innovations, structures, and airplanes. The base was formally dedicated on July 3, 1940, during the same week that the Tacoma Narrows Bridge opened. The week was full of ceremonies and parades in honor of the fact that supplies and personnel could be easily moved between McChord Field, Fort Lewis, and the

Puget Sound Navy Shipyard in Bremerton, Washington, in preparation for the nation's defense. The first airplanes to arrive at McChord Field were B-18As and B-23s, which were a part of the 17th Bombardment Group and the 89th Reconnaissance Squadron.

The time to celebrate came to an end with the attack on Pearl Harbor on December 7, 1941. McChord Field's mission changed quickly, from training, aerial resupply, and defense of the Pacific Northwest, to strictly defense. Satellite fields were quickly built at Port Angeles, Olympia, Kitsap, and Ephrats, all to support the increase in airplanes arriving at the base.

Early in the war years, at least 10 members of the 17th Bombardment Group were selected to participate in the Doolittle raid over Tokyo, flying B-25s off the aircraft carrier *Hornet*. The feat, never before attempted, shook the Japanese, who were surprised to see US aircraft flying over their homeland. As the war progressed, the mission at McChord expanded to include modification of P-39s and, later, redeployment of troops.

After World War II, McChord Field established itself as an airlift facility with the arrival in June 1947 of the 62nd Troop Carrier Group from Bergstrom Field, Texas. This group flew the Fairchild C-82A Packet, marking the beginning of a longtime association between the base and military airlift. On January 1, 1948, McChord Field was renamed McChord Air Force Base. As the decades progressed, the 62nd name underwent a few name changes, from the Troop Carrier Group, to the Military Airlift Wing, and finally the 62nd Airlift Wing. The nature of the airlift aircraft also changed, and included the C-82, C-54, C-124, C-141 and the C-17, to name a few. The command structure changed to GHQ Air Force Northwest Air District, to Military Air Transport Service, then to Military Airlift Command in 1966 and, finally, Air Mobility Command in 1992, as well as other tenant commands.

McChord's strategic location saw the P-61 Black Widow night fighters arrive as part of the 425th Night Fighter Squadron (NFS) from March Field, California, on September 1, 1946. The 425th NFS was the first fighter squadron assigned to the Air Defense Command. Later, the 317th Fighter Squadron replaced the 425th. The late 1940s and the 1950s saw the deactivation and reactivation of the 325th Fighter Interceptor Group/Wing, which encompassed the 317th and 318th Fighter Inspector Squadrons at McChord. As technology changed, so did the aircraft that defended the skies. Fighter aircraft transitioned from the 1940s through the 1980s, with the P-43, P-38, P-61, F-82, to the P-82, F-94, F-86D, F-102A, F-106, and F-15. These aircraft all played a role in the 25th Air Division's mission to protect the Western United States.

The photographs in this book tell the story of McChord Field, covering several decades of humanitarian airlift, defense of the Western United States, war, and numerous other conflicts around the around. Countless documents and historical reports, complied by the wing historians, tell the stories of those who sacrificed so much. It is with regret that not every name or event that took place in the wing's history could be mentioned in these pages. But this book will introduce you, the reader, to a few memorable events, faces, and aircraft that took place within the gates of McChord Field and that played a part in the air defense of the nation.

One

Building Blocks

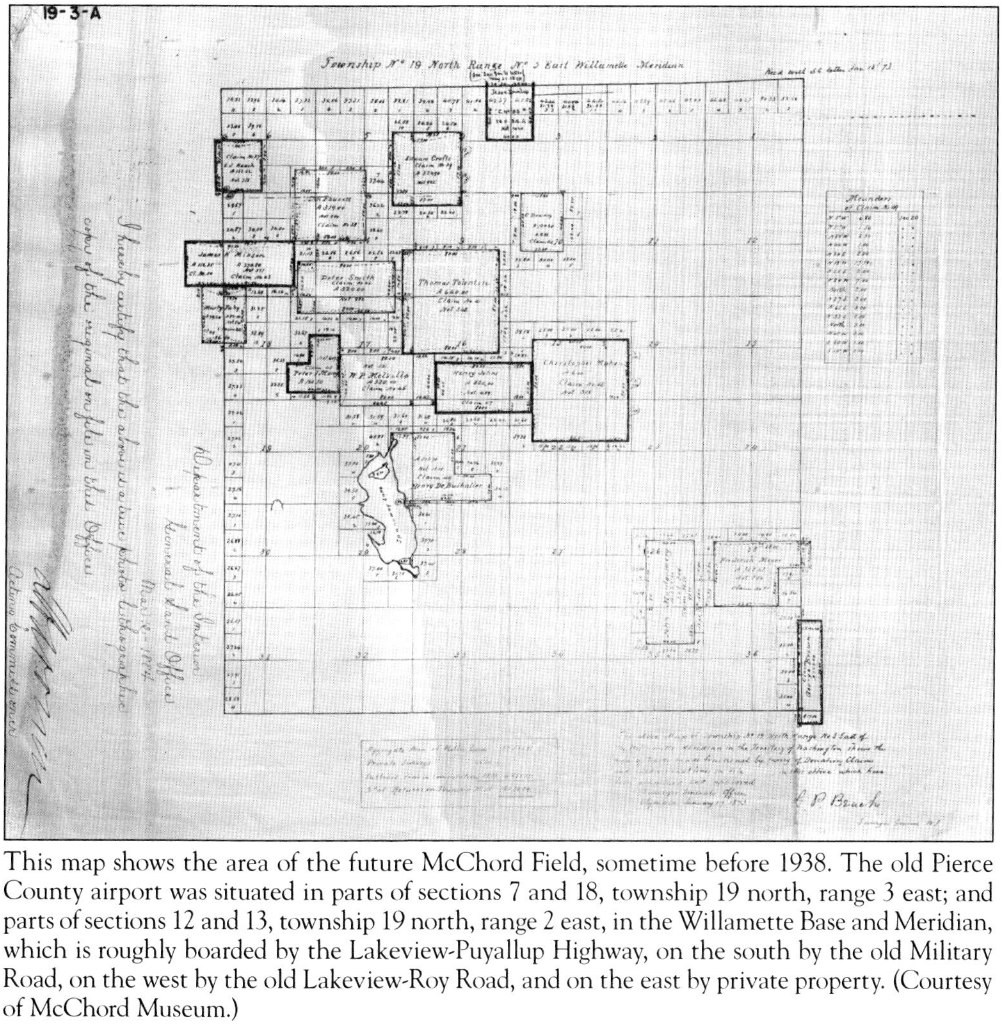

This map shows the area of the future McChord Field, sometime before 1938. The old Pierce County airport was situated in parts of sections 7 and 18, township 19 north, range 3 east; and parts of sections 12 and 13, township 19 north, range 2 east, in the Willamette Base and Meridian, which is roughly boarded by the Lakeview-Puyallup Highway, on the south by the old Military Road, on the west by the old Lakeview-Roy Road, and on the east by private property. (Courtesy of McChord Museum.)

Harold Bromley (right) poses with a man who is believed to be John Buffelen, Bromley's financial donor, under the *City of Tacoma*, a Lockheed Vega monoplane. The orange-colored craft had a 900-gallon fuel tank. Bromely used this plane in his attempt to fly solo and nonstop from Tacoma Field to Tokyo, Japan, a total of, 4,700 miles. This photograph was taken around 1929. (Courtesy of Tacoma Public Library.)

On July 28, 1929, Bromley made history, not by flying solo to Tokyo, but instead as the pilot of the first aircraft to crash at Tacoma Field. During takeoff, a gasoline tank sprayed fuel across the windshield. As Bromley went to look further, fuel was sprayed into his eyes, resulting in the plane veering off the runway and crashing. (Courtesy of Tacoma Public Library.)

In this May 1931 photograph, the large hangar has the name "Tacoma Field" written over the hangar doors. Tacoma Field was a logical choice for the federal government when it considered developing an air base to provide strategic air defense to the Puget Sound area. The new air base was called "the Northwest Air Base." (Courtesy of Tacoma Public Library.)

On March 30, 1938, Tacoma Field comprised 912 acres, a large hangar, and a 1,500-foot northeast-southeast runway. In this photograph, the deed passes between John Schlarb, chairman of the Board of County Commissioners (seated, left), and S. Clifford Davis, county auditor (standing, second from left). Col. Frank W. Wright, commander of the new air base, is standing on the far left. (Courtesy of Tacoma Public Library.)

The initial construction of McChord Field began in the summer of 1938. Workers from the Works Project Administration (WPA) began to clear the land, made up mostly of swamp, plains, and prairie. During construction, on a daily basis, the base employed 2,000 men, such as the WPA worker shown here, standing in a field of tree stumps that will become the site of Hangar No. 1. (Courtesy of McChord Museum.)

The three officers responsible for the construction of McChord Field stand under the Construction Quartermaster (CQM) sign. Shown here in February 1939 are, from left to right, James B. True (chief engineer), Lt. Col. Frank W. Wright (commanding officer), and Capt. E.P. Antonovich (construction quartermaster). During construction, the executive officers set up a room in the original hangar. (Courtesy of Tacoma Public Library.)

Sound Construction & Engineering of Seattle was awarded the contract for Hangars No. 1, 2, 3, and 4. Construction began in August 1938, and the hangars were completed in the fall of 1939, at a cost of more than $750,000. The hangars were built in two groups, each one resting on a concrete foundation measuring 253 feet, 6 inches by 290 feet, for a total of 73,515 square feet. Each hangar has offices, laboratories, shops, and storage rooms at each end. On top of Hangar No. 3 were a control tower, camera obscura, and meteorological station. In front of the hangars was a concrete warming apron that was approximately 600 feet wide and 2,600 feet long. The above photograph illustrates the initial construction of Hangar No. 1 on December 20, 1938. (Above, courtesy of Air Force Historical Research Agency; below, courtesy of Tacoma Public Library.)

On December 28, 1938, Ross B. Hammand & Company of Portland, Oregon, won the contract to construct the Air Corps barracks No. 100, pictured above. The structure, erected at a cost of $884,335.00, would billet between 1,285 and 1,300 men. A feature of building 100 was its consolidated mess hall, which could feed the entire command in one setting. (Courtesy of McChord Museum.)

In June 1939, contracts were awarded to MacDonald Building Company for construction of the officers' quarters (pictured) and to Howard S. Wright & Company, Inc., for construction of the noncommissioned officers' quarters. The floor plan called for a three-bedroom house with space for a maid's quarters in the basement. (Courtesy of Tacoma Public Library.)

It was not until October 29, 1941, that the base chapel held its first service inside the chapel. Religious services had been held in the Tacoma Hangar at the 17th Bombardment Group Day Room. According to the *Rip Chord* of February 21, 1941, Chaplain Cordill spoke about "Faith Founded a Nation" with the message that America was founded and preserved by faith and to help preserve America for future generations by keeping faith alive. (Courtesy of McChord Museum.)

The warehouses were completed on June 10, 1939, by Allen & Earley. Each building is identical in size, measuring 208 feet, 4 inches by 66 feet, 4 inches. A loading platform on each structure measured 16 feet by 8 feet on each end. (Courtesy of Air Force Historical Research Agency.)

In August 1939, preparation was under way by WPA crews to dig and reroute Clover Creek. Workers used gravel, dirt, and an 80-ton dragline to excavate a 20-foot-wide by 5-foot-deep channel two miles long. Before this work, the creek flowed in front of the hangars, separating them from the landing field. (Courtesy of Air Force Historical Research Agency.)

The radio transmitter building was completed on October 13, 1939. According to the base history, "The purpose of the building was to house an airdrome control transmitter (BC-329-A) and a point-to-point transmitter." The building's overall dimensions are 348 feet by 83 feet, 4 inches. It cost $40,703.77 to build. (Courtesy of Air Force Historical Research Agency.)

McChord Field is named in honor of Col. William C. McChord, who was born on December 29, 1881 in Lebanon, Kentucky. He graduated from the United States Military Academy on June 14, 1907, as a second lieutenant in the cavalry. Before attending the academy, McChord had earned a bachelor of law degree in 1901 from the Central University of Kentucky. (Courtesy of McChord Museum.)

In a letter dated May 10, 1908, 2nd Lieutenant McChord (pictured below) is reprimanded by the commanding officer at Camp Stotsenburg, Philippines, for riding a horse into the officers' clubhouse and into his quarters. The letter warns Lieutenant McChord that if he does this again, he will be arrested. (Courtesy of McChord Museum.)

Tragedy struck on August 18, 1937, when Colonel McChord, piloting an A-17 attack plane from Bolling Field, Washington, DC, to Randolph Field, Texas, died while trying to make a forced landing at Maidens, Virginia. According to McChord's biography, "The untimely and tragic death of Col. McChord was a distinct loss to the Air Corps. Those who knew him well realized his ability and . . . his initiative, independent thought and frank expression of opinion which characterized his service and which led his brother officers to rely so much upon his experience and helpful advice." During his active-duty career, Colonel McChord had a total of 3,000 hours as a pilot and observer. To keep his name alive, Tacoma Field, which had been renamed Northwest Air Base, was renamed McChord Field in honor of this courageous aviator. (Courtesy of McChord Museum.)

In March 1939, the drilling of two deep wells was completed. Two well turbines were installed, with the south well located about 90 feet west of B Street and 50 feet south of Fifth Street. The installation for the well turbines was not completed until April 8, 1940. The two deep wells had a capacity of 500 gallons per minute. (Courtesy of Air Force Historical Research Agency.)

This aerial photograph, taken on January 28, 1940, shows the Air Corps barracks in the foreground. The central heating plant, on the left side of the barracks, was built by Walter S. Leland, & Co., of San Francisco, California. The plant was designed with an ultimate capacity of 3,000 horsepower and provided heat for the hangars, barracks, hospital, and other structures. The heating facility used coal until the 1970s, when it was converted to natural gas. On the far left side edge of the photograph, the water tank stands tall over the landscape. The contract to construct the 300,000-gallon tower was awarded to the Chicago Bridge & Iron Company of San Francisco, California, on June 10, 1939, The water tower, used to provide a reserve water supply, was not completed until June 1940. (Courtesy of Air Force Historical Research Agency.)

On June 19, 1939, the hospital was completed. The building is two stories and measures 38 feet, 4 inches by 88 feet, 4 inches, and could accommodate 19 patients and eight attendants. At left can be seen a three-stall ambulance garage. By the end of 1939, the base had 12 officers and 88 enlisted men, but it was not until the end of December 1941 that four nurses arrived at McChord Field. (Courtesy of Air Force Historical Research Agency.)

Runway construction began in November 1939 and was completed in 1940. The strips had a load-bearing capacity of 12,000 pounds. The first and second runways were each 5,865 feet long and 150 feet wide, and the third was 5,000 feet long and 150 feet wide. This made it possible for the 17th Bombardment Group and the 89th Reconnaissance Squadron to arrive in June 1940 with Douglas B-18s, B-18As, and B-23s, as pictured above. The image below, taken on June 25, 1940, shows the same aircraft and a closer view of Hangars 1 and 2, with Mt. Rainier in the background. (Above, courtesy of McChord Museum; below, courtesy of Tacoma Public Library.)

The Boeing B-17B Flying Fortress is on static display during McChord Field's dedication ceremony on July 3, 1940. In the afternoon, the public as watched 100 Army planes, such as the B-17B, put on a flying demonstration. An interesting note is that the B-17C was used to train 40 Royal Air Force aircrews from No. 90 Squadron from January thru April 1941. (Courtesy of McChord Museum.)

Seen here is another angle taken on July 3, 1940, at McChord Field's dedication ceremony. A large crowd has gathered in front of the hangars with B-18 and B-23 aircraft parked on the ramp, and a dirigible is flying in the background. The Gilmore loudspeakers and ambulance are in the foreground to the right of the photograph. (Courtesy of Tacoma Public Library.)

This photograph was taken on November 25, 1940, of new recruits receiving training at the new Recruit Center on McChord Field. They will have six weeks of training, including the use of a gas mask. Note the two recruits in the center of the image with their gas masks on during a demonstration of the proper use of the gas mask. (Courtesy of Tacoma Public Library.)

This is another photograph of the 900 new recruits on McChord Field, this time marching down the flight line, in November 1940. Each recruit is destined for specialized branches of service in the medical corps, signal corps, and ordnance. At the time this photograph was taken, each recruit was supposed to serve on active duty for three years. (Courtesy of Tacoma Public Library.)

Two

Defend, Train, Modify, and Redeploy

At the end of 1941, according to the finance department, about 4,000 enlisted men, 200–250 officers, four nurses, and two warrant officers were assigned to McChord Field. The base's airplanes included B-18s, B-18As, and B-23s. In this photograph, four B-18A Bolo Bombers assigned to the 17th Bombardment Group (Medium) sit on the ramp, with Mount Rainer in the background. (Courtesy of Ernest White.)

After the attack on Pearl Harbor, McChord Field went from routine maintenance of keeping the field in efficient operations to the feverish activity of a nation at war. Planes and men from the 17th Bombardment Group began patrolling the nearby coastline for Japanese submarines. The 17th Bombardment Group traded in their B-18s, B-18A, and B-23s to become the first unit to receive the North American B-25 Mitchell Medium Bomber. (Courtesy of McChord Museum.)

One of the pilots that flew a B-25 and participated in the Doolittle raid was 1st Lt. Everett W. Holstrom from Westter, Oregon. He arrived at McChord Field on September 28, 1940. After the raid over Tokyo, Lieutenant Holstrom received the Distinguished Service Cross and later became a brigadier general, retiring from active duty on July 1, 1969. (Courtesy of Air Force Historical Research Agency.)

Another member of General Doolittle's raiders was 1st Lt. Edward J. York. He was born on August 18, 1912, and attended high school in Batavia, New York. He arrived at McChord Field on June 29, 1940. After the raid on Tokyo, he received the Distinguished Service Cross. (Courtesy of Air Force Historical Research Agency.)

1st Lt. David M. Jones was another member of the Doolittle raid who received the Distinguished Service Cross. He was born on December 18, 1913, and called Charleston, Oregon, home. He joined the Reserve Officer Training Corps (ROTC) program at the University of Arizona, and arrived at McChord Field on June 29, 1940. (Courtesy of Air Force Historical Research Agency.)

Capt. Ted W. Lawson was from Los Angeles, California. He attended Alan Hannock College of Aeronautics and flying school at Randolph and Kelly Fields, San Antonio, Texas. Captain Lawson lost his leg on April 18, 1942, and later wrote a book, *Thirty Seconds Over Tokyo*, about the Doolittle raid. He, too, was awarded the Distinguished Service Cross for his participation in that mission. (Courtesy of Air Force Historical Research Agency.)

Another member of the Doolittle raid, also awarded the Distinguished Service Cross, was Maj. Charles Ross Greening. He arrived at McChord Field in June 1940. Greening served as the public relations officer and was appointed the first base photographic officer. Major Greening designed what was known as the "twenty cent bombsight." This piece of equipment replaced the Norden bombsight, which proved to be ineffective at low attitudes. The name came from the fact that the materials used in the bombsight cost only 20¢. Greening's story of the raid over Tokyo can be found in McChord Field's publication, *Rip-Chord* (Vol. 4, No. 1, July 4, 1942, page 2). (Courtesy of Air Force Historical Research Agency.)

1st Lt. Harold F. Watson went through the ROTC program at Norwich University and Virginia Military Institute, then attended Parks Air College and Randolph and Kelly Fields in Texas. He arrived at McChord on September 28, 1940. Like other members of the Doolittle raid, he was a part of the 17th Bombardment Group, which returned from Pendleton, Oregon, to McChord Field after the bombing of Pearl Harbor. He was awarded the Distinguished Service Cross. (Courtesy of Air Force Historical Research Agency.)

Capt. John A. Hilger reported to McChord on June 2, 1940, and was appointed base photographic officer in September 1940. He, too, received the Distinguished Service Cross for his efforts during the Doolittle raid. Hilger attained the rank of brigadier general. He retired from active-duty service on November 30, 1966. (Courtesy of Air Force Historical Research Agency.)

An article published in the *Tacoma Times* on February 23, 1944, titled "K-9 Corps Guards McChord Field" explained the major importance of the K-9 Corps, members of which worked closely with their trainers, guarding 2,500 acres of McChord Field at night against "nocturnal intruders." Besides guarding the base, the dogs' other duties included carrying messages from one post to another. The article also praises the dogs for obeying their trainers and their ability to work at great distances from their trainers by responding to a silent whistle. Lastly, the dogs received the best medical attention under the supervision of the Veterinary Department in 1943–1944. (Courtesy of Air Force Historical Research Agency.)

Even during a war, base life continued. Helping to relieve a little stress, members watched a boxing match between Paine Field and McChord Field at McChord's field house sometime in 1943–1944. The names of the boxers and referee are unknown. (Courtesy of Air Force Historical Research Agency.)

Members of McChord Field's bowling team, the McChordites, led the North End Classic League of Tacoma at the end of 1944 and early 1945. The team held most league records for individual and team scores. Pictured from left to right are (crouching) Pvt. Ed Feuerstein and T.Sgt. LeRoy Willin; (standing) Pvt. Arthur Pence, CWO Donald Tippie, and Sgt. Ed Novak. (Courtesy of Air Force Historical Research Agency.)

On March 15, 1944, Capt. Maude Kennedy Schaub was the Women's Army Corps (WAC) commanding officer, WAC department. She arrived at McChord Field on May 30, 1943, and was part of the 771st WAAC Post Headquarters Company that was activated on May 24, 1943, per G.O. No. 105, Headquarters Fourth Air Force, as a first lieutenant but was promoted to captain on September 1, 1943. (Courtesy of Air Force Research Historical Agency.)

Squadron W, composed of WAC personnel, had 74 women assigned to McChord at the beginning of October 1944. By the end of the month, there were 86 WACs stationed at the base. The increase in numbers was due to the processing activities on the base to redeploy members of the 11th Air Force from Alaska. Shown here is 1st Lt. Carol S. Goodman, commanding officer, WAC detachment. (Courtesy of Air Force Historical Research Agency.)

Members of Squadron "W," WACs, now have a homey atmosphere to socialize and dance with their dates in their newly converted Date Room, pictured here on January 8, 1945. The room formerly was the squadron's former mess hall, but now the ladies and their friends could purchase beer and sandwiches during the evening. According to an article in the *Rip-Chord,* January 13, 1945, once the room was completely decorated it would have been the most attractive and relaxing centers on McChord Firled. (Courtesy of Air Force Historical Research Agency.)

This photograph of Butch, the WAC mascot, appeared in the *Rip-Chord* on January 27, 1945, wearing his uniform and stripes for the last time. The articler described Butch as having an indeterminate ancestry. He was an overzealous protector of his charges and was sent home with Cpl. Jan Wallace to Iowa. (Courtesy of Air Force Historical Research Agency.)

A call "To the Colors" is sounded during a formal retreat review on October 19, 1944. Col. Theodore M. Bolen made a presentation of two Air Medals. Standing at the front of the review are, from left to right, WAC officer Lt. Jane S. McILroy, Henry Mamock, and Mrs. Cecil Parker. Lt. Henry E. Mamlock and Sgt. Jack A. Parker were held as prisoners of war during a combat mission over occupied Europe. (Courtesy of Air Force Historical Research Agency.)

Henry Mamlock (center), of Tacoma, receives the Air Medal with two oak-leaf clusters on behalf of his son, 2nd Lt. Henry E. Mamlock, from Colonel Bolen. Mrs. Cecil Parker (second from right), from Auburn, also accepted the Air Medal, on behalf of her son, Sgt. Jack A. Parker. WAC officer Lt. Jane McILroy, public relations officer, stands at left. (Courtesy of Air Force Historical Research Agency.)

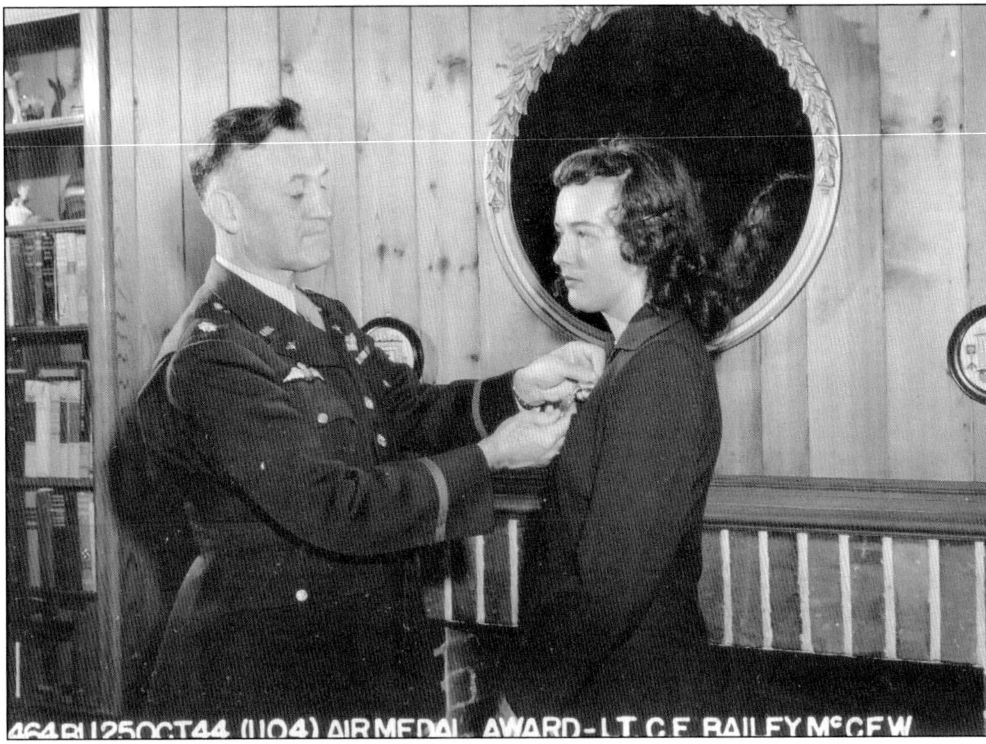

Lt. Col. Frank A. Flynn presents the Air Medal with one oak-leaf cluster to Janet A. Bailey on October 25, 1944, at the Bailey residence. Janet's husband, 2nd Lt. Carl M. Bailey, Air Corps, was declared missing in action. (Courtesy of Air Force Historical Research Agency.)

This photograph appeared in the *Rip-Chord* of March 10, 1945. The caption reads, "The little son of Major Theodor Willhite [a B-24 crew member serving in Italy], who will never remember his father, here receives the Distinguished Flying Cross, posthumously awarded Major Willhite for gallantry and achievement in action. Col Theodore M. Bolen, base commanding officer, presented the medal." (Courtesy of Air Force Research Historical Agency.)

Pilots ferried bombers from Seattle to McChord, where the aircraft awaited Ferry Command pilots to take them to various destinations. The Ferry Command at the base also became responsible for ferrying the Bell P-39 Airacobras to their final destinations—before the P-39 was put into storage at McChord in the fall of 1945. (Courtesy of Air Force Historical Research Agency.)

Even in war, life went on. Shown here is the wedding of Pfc. Mary Zander and S.Sgt. Alec J. Demanovich. (Courtesy of Air Force Historical Research Agency.)

Standing in front of the post chapel and smiling happily are newlyweds Pvt. Ruth Beckwith Krug of Mobridge, South Dakota, and Pfc. Horst Krug of Toledo, Ohio. The happy couple was married on November 4, 1944. (Courtesy of Air Force Historical Research Agency.)

One featured section in the *Rip-Chord* of late 1944 and early 1945 was called "Around the WAC Shack." Topics varied from announcements to gossip, such as the report of an engagement of a G.I. temporarily stationed at Fort Lewis. As pictured here, it was not gossip. Cato and Lawson were married on February 25, 1944. (Courtesy of Air Force Historical Research Agency.)

Shown here are scenes from the Christmas Eve party at the officers' club on December 24, 1944. (Courtesy of Air Force Historical Research Agency.)

This is the reception line for the commanding officer's annual New Year's Day reception. Shown here in 1945 are, from left to right, Capt. C.D. Whitehead (adjutant), Col E.C. French (executive officer), Col. Theodore M. Bolen (commanding officer of McChord Field), Mrs. T.M. Bolen, Mrs. E.C. French, and Mrs. C.D. Whitehead. (Courtesy of Air Force Historical Research Agency.)

The refreshments table is well stocked at the commanding officer's annual New Year's Day reception, January 1, 1945. (Courtesy of Air Force Historical Research Agency.)

McChord Field faced a new challenge of redeployment of troops from the 11th Air Force in Alaska. The base received notification on October 19, 1944. Six days later, on October 25, a total of 184 men arrived from Unit 4. Shortly afterwards, more troops arrived from Units 5 and 6, 310 Fighter Squadron, and from the 1741st Ordnance Supply and Maintenance Company. (Courtesy of Air Force Historical Agency.)

The members of the 11th Air Force returned from their overseas duty in Alaska in October 1944. Their first point of entry was McChord Field. Records were checked, pay was brought up to date, old uniforms were turned in and new ones issued, physicals were given, and transportation was arranged for furlough travel under the Army's new Rehabilitation, Recuperation and Recovery Plan. (Courtesy of Air Force Historical Research Agency.)

During processing, one of the biggest hurdles was finding enough typists to handle all of the necessary paperwork. In addition, personnel had to be provided to assist in various locations along the processing line. Additional military and civilian personnel came in for a week to help, but had to return to their main duty stations. (Courtesy of Air Force Historical Agency.)

Despite the short notification, McChord Field performed with style and speed, which was greatly appreciated by the men returning from duty in Alaska and experiencing their first homecoming. Unit 4 arrived at 10:30 a.m., and the last man was passed through the line at 1:45 p.m. This soldier has just cleared all of his processing and now begins his 30-day furlough. (Courtesy of Air Force Historical Research Agency.)

Troops from Alaska went to building 7, one of seven sites set up to handle the processing line. There, the McChord Field Women's Club, made up of wives and widows of officers, sewed insignia, patches, and made other adjustments to uniforms. In this photograph, Mrs. Bolen, wife of Col. Theodore Bolen, sews an overseas stripe on the sleeve of a retuning GI. (Courtesy of Air Force Historical Research Agency.)

The McChord Women's Club was the last stop in the processing line after stops at records, medical, quartermaster, final, intelligence (briefing on what to say and what not say while on leave), Squadron D Detachment Headquarters Information, and final clearance. The women in this photograph provided their sewing services free of charge to the troops of the 11th Air Force in late October 1944. (Courtesy of Air Force Historical Research Agency.)

The president of the United States designated October 8–14, 1944, National Fire Prevention Week, and the base actively participated. A "Fire Fighting Equipment Display" was set up in the lobby of building 100. The display featured an asbestos firefighter's suit and several types of fire extinguishers, ranging in size from one quart to 50 pounds, seen on the table behind the two enlisted WACs. (Courtesy of Air Force Historical Research Agency.)

National Fire Prevention Week was promoted on calendar displays around the base. The month of October had a different color background, and the slogan "To speed victory, prevent fires in the Army Air Forces" was displayed on the calendar. In addition, a firefighting demonstration was held, as shown here. (Courtesy of Air Force Historical Research Agency.)

The Special Services Department of McChord Field, whose mission was to aid in the morale of the base's personnel, teamed up with the Convalescent Training Program to create a broadcast system running from the post theater to the base hospital. On October 17, 1944, Sgt. I. Leowiz (left), Corporal Crawford (center), and Sergeant Bousman made this a reality by broadcasting a USO show to the hospital patients. (Courtesy of Air Force Historical Research Agency.)

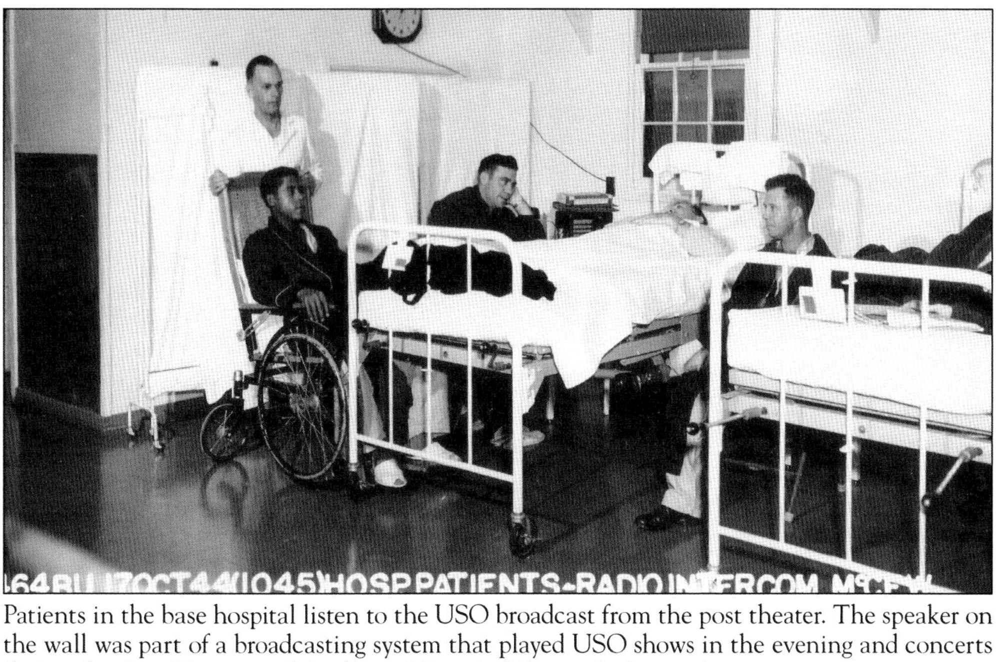

Patients in the base hospital listen to the USO broadcast from the post theater. The speaker on the wall was part of a broadcasting system that played USO shows in the evening and concerts during the day. (Courtesy of Air Force Historical Research Agency.)

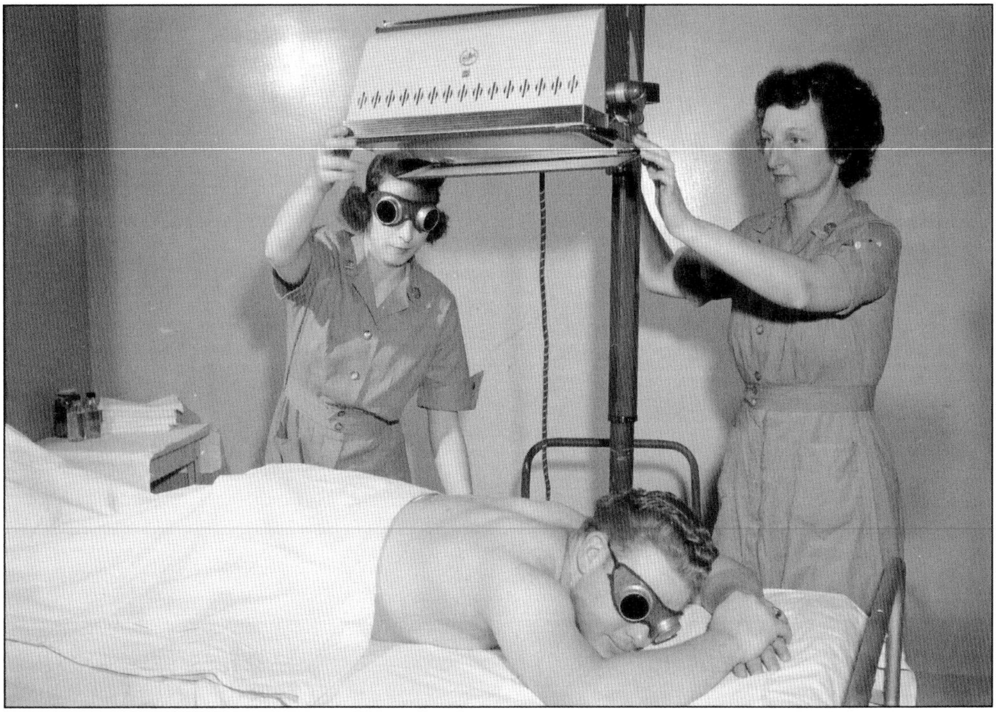

On October 17, 1944, ultraviolet ray treatment is given to a GI by two WACs in the hospital. (Courtesy of Air Force Historical Research Agency.)

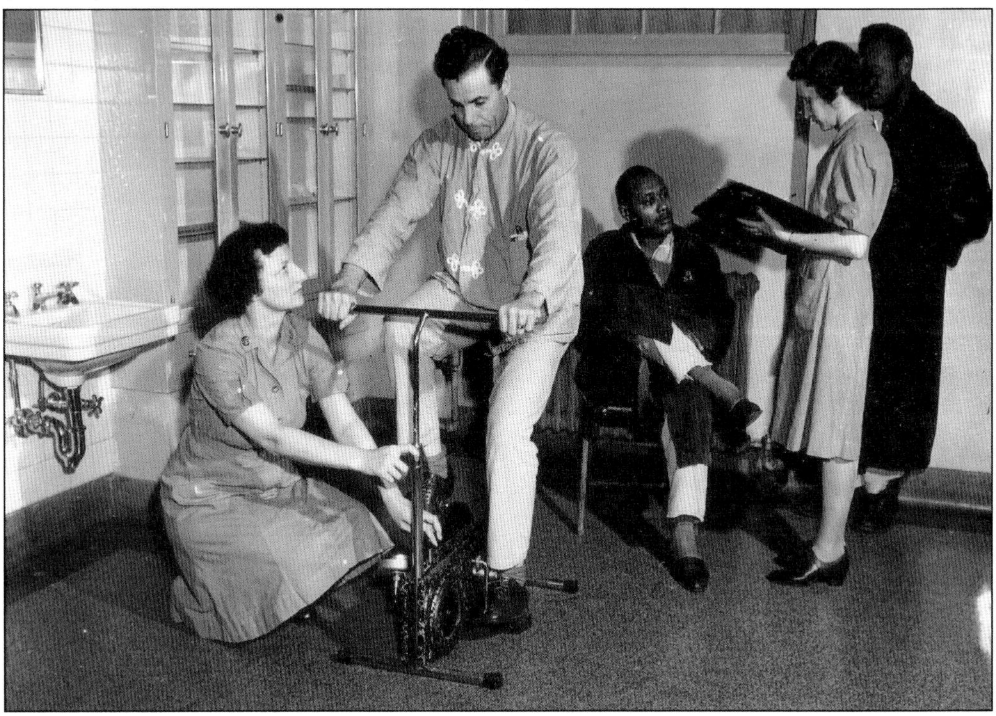

WACs at the base hospital are seen here helping soldiers in the physiotherapy ward. (Courtesy of Air Force Historical Research Agency.)

Another mission of McChord Field was to serve as a modification and repair center for the Bell P-39 Airacobras. The Manning Section, 4th Air Force, authorized the employment of an additional 300 civilians and 40 enlisted men. Arrangements were made with Clover Park Vocational School to provide instruction for training the new employees in skills related to aircraft maintenance. Trainees were paid $1,500 a year. In June 1944, 125 out of 140 workers at the modification center had been trained at Clover Park. Capt. Mabon A. Cornwell was in charge of setting up the modification center. He had help from experts at Bell Aircraft Co. and Allison Motors. (Above, courtesy of McChord Museum; below, courtesy of Air Force Research Historical Agency.)

One of the major concerns while setting up the modification center was hiring enough civilians, because the civilian agencies had higher paid wages and the Army Air Forces could not compete. The manpower shortage became more of a problem when the modification program increased from 10 to 25 planes per month. The top photograph was taken on October 2, 1944, inside one of the main hangars on McChord Field. A post exchange restaurant was established in Hangars 1 and 2 for employees of the modification center to purchase refreshments during their break. The photograph below is of the last P-39 to be modified at McChord Field on Wednesday, September 27, 1944. (Above, courtesy of Air Force Research Historical Agency; below, courtesy of McChord Museum.)

The Infantry Day Parade, held in conjunction with the Pierce County War Bond Drive, took place on June 15, 1944. Civilian employees of McChord Field's modification center rode in a float, working on an aircraft engine. Earlier in the week, on June 11, 1944, the base participated in Tacoma's Fifth War Loan Drive by having an open house. About 15,000 people came to view the presentation of awards and to buy war bonds. (Courtesy of Air Force Historical Research Agency.)

This is one of four floats in the June 1944 Infantry Day Parade that depicted activities at McChord Field. On this float, a P-39 fires blank ammunition at a Japanese Zeke. This demonstration of the P-39's power was greatly enjoyed by the parade's onlookers. Approximately 250 McChord Field troops marched in the parade, including WACs and enlisted men. (Courtesy of Air Force Historical Research Agency.)

On Flag Day, June 14, 1944, a retreat parade and review of troops was held. Here, McChord officers salute in front of building 100, as the review of colors are presented to Col. Theodore M. Bolen and his staff. The presentation of colors included the flags of Canada, Great Britain, Russia, and the 4th Air Force flag. (Courtesy of Air Force Historical Research Agency.)

On September 14, 1944, Good Conduct Awards were presented to Squadron A for its members' soldierly qualities, at an impressive parade ceremony. Here, Col. Theodore M. Bolen, commanding officer, ties blue and gold streamers on the Squadron M, Medical Detachment, guidon in recognition of its excellence in drill and appearance. (Courtesy of Air Force Historical Research Agency.)

On June 29, 1944, officials from Army Air Force, Washington, DC, arrived and inspected McChord Field at 3:00 p.m. In a large mass presentation, Good Conduct Medals were awarded to 98 members of Squadron C. Receiving his award from Colonel Bolen, is 1st Sgt. Phillip G. White. Behind Sergeant White, at right, is Lt. Col. C.F. Rugh presenting awards to other members of Squadron C. (Courtesy of Air Force Historical Research Agency.)

On June 22, 1944, Cpl. Clarence B. Clement (right) a member of the Army Air Force Communication Service, receives the congratulations of Capt. Charles D. Whitehead, McChord Field's base adjutant. Clement was being awarded the Soldier's Medal for heroism in the Alaskan theater. (Courtesy of Air Force Historical Research Agency.)

Capt. T.M. Bolen had been assigned to McChord Field on March 19, 1940. As a captain, his primary duty was as post engineering officer. Among his other roles were post adjutant, post operations officer, public relations officer, and post exchange officer. Almost four years later, Captain Bolen, now Colonel Bolen, returned as base commander under General Order No. 5, AB Hq. (Courtesy of Air Force Historical Research Agency.)

In an effort to show the good relationship between Fort Lewis and McChord Field, a group of Fort Lewis engineer troops from the 72nd Engineer Battalion, in search of training activities and undergoing their basic phase of engineering training, would assist with repairing roads and bridges, filling in gun emplacements, removing revetments, and building an emergency flight strip southwest of the field. On September 6, 1944, the Fort Lewis engineers arrived and set up a field mess, seen in the above photograph. At the end of the day, the troops enjoyed a field shower (left). (Courtesy of Air Force Historical Research Agency.)

In an effort to show its appreciation to the Fort Lewis engineers, the Special Services Department at McChord Field entertained boys of the 73rd Engineer Battalion with singing, dancing, and other performances in September 1944. Here, Cpl. Claude W. Frankfurter, 577th AAF Band, beats out rhythm for the Fort Lewis troops. (Courtesy of Air Force Historical Research Agency.)

Shown here is another performance by the Special Services Department in September 1944 for the 72nd Engineer Battalion. One of the enlisted men has a "Come to Papa" look as he plays his part with Ruth Michaels in one of the several performances offered. (Courtesy of Air Force Historical Research Agency.)

August 21, 1944, marked the first arrival of an ambulance plane, a C-47, which brought 12 patients from the Alaskan Theater, 11th Air Force, to McChord Field. Air Transport Command (ATC) had developed two transport routes to Alaska to support the 11th Air Force. One route was from McChord Field, and the other was developed to support the Aleutian campaign. The patients that were transferred to the Fort Lewis hospital from the C-47 may have been wounded from the Aleutian campaign. According to the base summary of August 1944, there would be more ambulance planes arriving from Alaska. (Courtesy of Air Force Historical Research Agency.)

The C-47 Skytrain, also known by the nickname "The Gooney Bird," was the backbone of the Army Air Force. It was deployed to every combat area during World War II. The C-47 provided air logistics, troop transport, airdrop, rescue, reconnaissance, glider towing, navigator training, and special operations—a role the C-47 played into the 1960s. The above photograph shows the interior section of a C-47 plane from Alaska, carrying patients. (Above, courtesy of Air Force Historical Research Agency; below, courtesy of McChord Museum.)

Three

Heavy Lifting and Jet Power

The Fairchild C-82 Packet arrived at McChord Field along with the 62nd Troop Carrier Group, on August 15, 1947, from Bergstrom Field, Texas. The C-82s flew at McChord until 1949 and with "United States Air Force" painted on the body, as seen in this photograph. They were involved in such missions as Project Yukon and Operation Hayride, and they flew flood-relief supplies to several locations in Washington and Oregon. (Courtesy of McChord Museum.)

The Northrop P-61 Black Widows were transferred from March Field, California, to McChord on September 1, 1946, as part of the 425th Night Fighter Squadron (NFS), making it the first squadron assigned to the Air Defense Command. The 425th NFS was deactivated and replaced by the 317th Fighter Squadron in August 1947. The 317th continued to fly the P-61 until the squadron was reassigned to Hamilton Field, California. (Courtesy of McChord Museum.)

This 1948 close-up photograph of the Boeing SB-17G, which was assigned to McChord Air Force Base as part of Flight C, 4th Air Rescue Squadron. Note the 27-foot lifeboat loaded with survival gear attached under the belly. Flight C's area of responsibility was Puget Sound, Washington, Oregon, Idaho, Montana, and the Pacific coast. The "SB" designation distinguishes the search-and-rescue mission from the bomber missions of World War II. (Courtesy of McChord Museum.)

This photograph, taken on May 8, 1946, shows Manuel Acuna Roxas, Philippines' president-elect (right), and US High Commissioner Paul McNutt (center) speaking with Col. A.S. Strickland, commander of McChord Field. The C-54 airplane in the background flew from Manila to Washington, DC, and stopped to fuel at McChord. The base was the halfway point of the trip, which started in Manila, then to Tokyo, and then through the Aleutians before arriving at McChord Field. Roxas went to Washington, DC, to meet with President Truman about an offer of military bases and support in the Philippines. Roxas and McNutt spent an hour at the officers' club, drinking coffee with personnel of the base, before departing at 9:30 a.m. (Courtesy of Tacoma Public Library.)

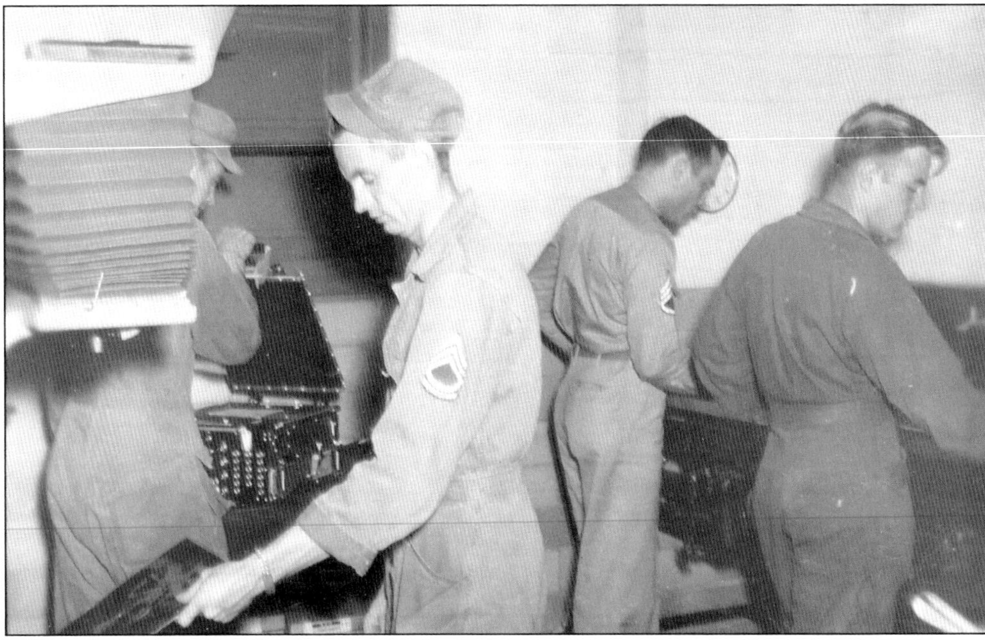

This September 1947 photograph shows the 62nd Airdrome Group Photostat Section, base laboratory. The photostat was a machine used to make photocopies on special paper. This section's duties were to reproduce legal documents, all classified material, and all reproductions. Shown here are, from left to right, S.Sgt. Charlie A. Tigrett, T.Sgt. John R. Potter, S.Sgt. Harry M. Hanst, and Pfc. Grover A. Salmon. (Courtesy of Air Force Historical Research Agency.)

Shown here is an earlier model of a copy machine. This large camera is a copying camera, used in the base photographic laboratory in September 1947. The copying camera had a 19-inch focal lens and a 20-by-24-inch negative back on which the user placed the original document to copy it. (Courtesy of Air Force Historical Research Agency.)

Personnel from the Shipping and Receiving Section, Commercial Transportation Section, are shown here in November 1947. M.Sgt. Edward R. Landis (left) and M.Sgt. Russel P. Matthew examined, handled, and reviewed paperwork for all incoming and outgoing freight from McChord Air Force Base. (Courtesy of Air Force Historical Research Agency.)

Personnel from the 62 Airdrome Commercial Transportation Section, Pfc. John L. Hummelguard (left) and M.Sgt. Adam Huray, prepare cargo to be shipped from McChord Air Force Base. The freight section was responsible for preparing the crate, weighing the cargo, and shipping it to various locations around the world. (Courtesy of Air Force Historical Research Agency.)

The Children's Christmas Party was held on December 21, 1947, at the base theater. The 62nd Airdrome Group, Special Services Section, sponsored the event. Two little girls receive presents from Santa Claus, who was played by M.Sgt. C.L. Childs of Special Services. (Courtesy of Air Force Historical Research Agency.)

Cpl. James D. Love, a member of Squadron F of the 62nd Airdrome Group, was assigned to the Base Signal Office on April 7, 1947. According to his Enlisted Performance Report, Corporal Love was one of six teletype operators on base with the Military Occupation Special 237. One of his job requirements was to operate the telegraphic typewriter for transmitting and receiving messages. (Courtesy of Air Force Historical Research Agency.)

Members of Squadron A, Telephone Switchboard Personnel of the 62nd Airdrome Group, take a moment to pose for a photograph on February 18, 1948. One of the many important functions performed by the telephone operator was to dispatch the ambulance, the fire department, and military police in the event of any emergency on the base. (Courtesy of Air Force Historical Research Agency.)

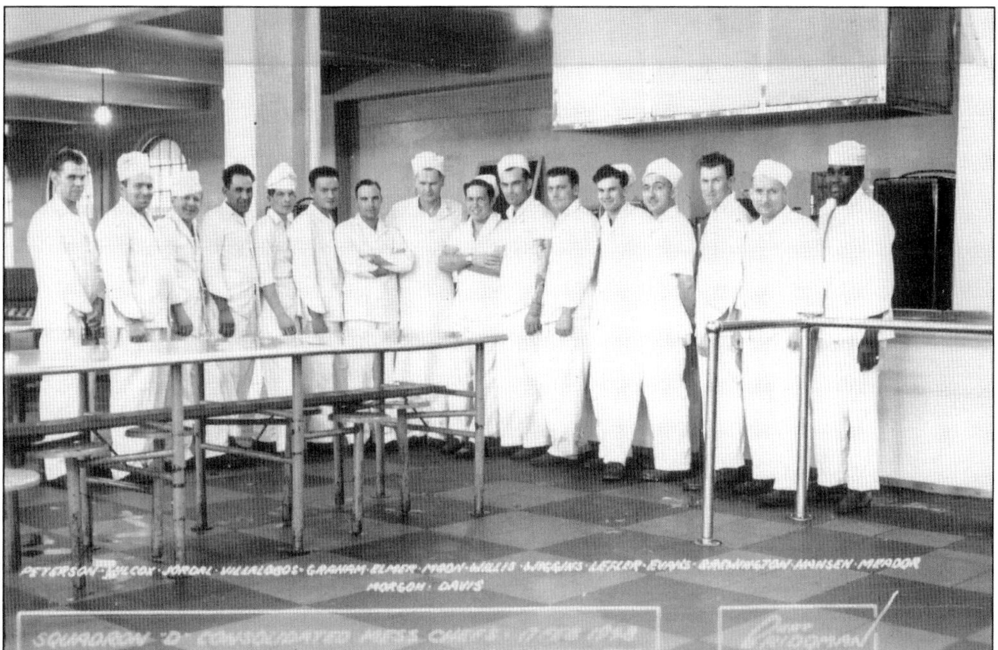

Anyone who has ever eaten in a dining facility appreciates the effort put forth by the cooks and chefs. Members of Squadron D, Consolidated Mess Chefs of the 62nd Airdrome Group, pose for a photograph on February 17, 1948. (Courtesy of Air Force Historical Research Agency.)

Members of Squadron E, Drivers School, 62nd Airdrome Group, pose on February 26, 1948. Posters attached to the counter read, " Stay Alive Obey Traffic Signals" and "Highway Habits," reminding drivers to remain safe on the road. (Courtesy of Air Force Historical Research Agency.)

This photograph from early winter 1948 shows the Fire Department Crash Personnel, 62 Airdrome Group. The group's mascot is sitting in the front row. (Courtesy of Air Force Historical Research Agency.)

This photograph was taken in the winter of 1949, during Operation Hayride. The 62nd Troop Carrier Group used the C-82 Packet. Hay bales, unloaded from a large trailer onto the C-82, were to be airdropped to stranded livestock, an estimated 4 million head of cattle and sheep, during an unusually harsh winter in late 1948 and early 1949. Due to the harsh conditions, ranchers could not reach their livestock overland. President Truman had C-82 units, such as the one at McChord Air Force Base, assist in efforts that helped save at least 80 percent of the livestock in the affected states of Arizona, Colorado, Kansas, Nebraska, Nevada, and Utah, an area of more than 193,000 square miles. (Both, courtesy of McChord Museum.)

After World War II, McChord Field, and later McChord Air Force Base and assigned assets, participated in various humanitarian missions, such as Operation Hayride and the Berlin Airlift. By the early 1950s, however, McChord found itself heavily involved in airlift operations involving C-54s and C-124s. McChord had established itself as an airlift facility and as a major port for moving troops and supplies during the Korean War. In these photographs from the late fall of 1950, soldiers wounded in the Korean conflict are offloaded and disembark on a converted passenger ramp. (Courtesy of Air Force Historical Research Agency.)

At the time, the Military Air Transport Service (MATS) maintained the longest air-supply route, moving essential cargo. Between June 1950 and July 1953, MATS flew 43,196 causalities from Japan to the United States. In the right photograph, the 62nd Troop Carrier Group (later the 1705th Air Transport Group, after the 62nd moved to Larson AFB) unload wounded from the Korean conflict in the late fall 1950. One of the aircraft used during the conflict, the C-54, can be seen parked on the ramp, facing Mount Rainier. (Courtesy of Air Force Historical Research Agency.)

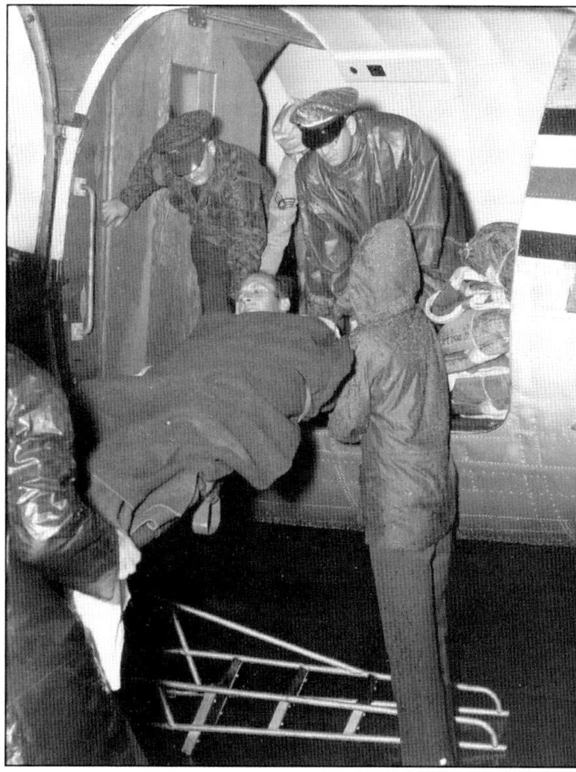

This photograph was taken in the early summer of 1951 on McChord Air Force Base, at the passenger terminal, or, as the sign reads, the MATS terminal. Note the different branches and service uniforms, including Royal Canadian Air Force (sergeant), US Air Force (technical sergeant), US Marine Corps (private first class), US Navy (unknown rank), and US Army (private). (Courtesy of Air Force Historical Research Agency.)

At the new 1705th Air Transport Wing cargo shed, freight is waiting to be loaded onto the Douglas North Star, the Canadian equivalent of the C-54. This particular plane is from the Royal Canadian Air Force's (RCAF) 426th Thunderbird Squadron, under Wing Cdr. C.H. Mussells. The RCAF assisted with supplies and troops from McChord AFB to Alaska and Japan. (Courtesy of Air Force Historical Research Agency.)

The Korean War started on June 25, 1950, and a month later, the RCAF 426th Thunderbird Squadron was deployed to McChord AFB. During the Korean conflict, while assigned to McChord, the 426th Squadron flew 599 round-trips to the Far East, a total of 34,000 flying hours, without a loss of cargo or passenger. In these photographs, the first contingent of UN Canadian troops boards an aircraft of the 1705th Air Transport Wing, Continental Division, Military Air Transport Service, for their flight to the Far East in January 1951. (Courtesy of Air Force Historical Research Agency.)

Early in the Korean conflict, MATS had delivered only 2.5 tons of cargo from the United States to Japan. By the time of the truce on July 27, 1953, MATS had moved 214,000 passengers and 80,000 tons of cargo to Japan, using C-47s, C-54s, C-97s, C-119s, and C-124s. These two photographs, taken on June 8, 1951, illustrate the large scale of troop movement from neighboring Fort Lewis. McChord AFB, which was part of MATS, moved these 200 Army troops onto a C-124 aircraft. (Courtesy of Air Force Historical Research Agency.)

Actor Keenan Wynn (right) and French actress Denise Darcel visit with an unidentified officer from McChord Air Force Base on November 5, 1951, at the officers' club dance. Wynn stared in popular movies such as *Dr. Strangelove* and *Kiss Me, Kate*, and Darcel appeared in *Tarzan* and *Battleground*. The troupe put on a 90-minute performance at McChord AFB. (Courtesy of Tacoma Public Library.)

Col. Richard Bromiley (second from left) served as commander of the 1705th Air Transport Wing from September 1950 to the middle of 1951. Here, he and members of the 1726th Support Squadron test food prepared in "hot cans," which were used like a thermos to keep food warm. Among the food tested were hamburgers, bread, butter, and fruit juice. (Courtesy of Air Force Historical Research Agency.)

Colonel Bromiley (center), commanding officer of the 1705th Air Transport Wing, congratulates Cpl. J. Gary of McChord AFB for his performance on the *MATS Television Show* on local station KING-TV, Seattle, which later became KING 5 TV. Standing at right is Captain Abner, who hosted the show on May 12, 1951. Note that Corporal Gary has two stripes on his arm and held the rank of corporal, a title of rank left over from the Army Air Force days. It was not until March 1952 and Air Force Regulation 39-36 that the titles of enlisted ranks were changed. (Courtesy of Air Force Historical Research Agency.)

In September 1951, the 25th Air Division Defense moved its headquarters from Silver Lake, Washington, to McChord AFB, and remained there until 1990. In the early 1950s, the 25th Air Division's main responsibility was defending the Pacific Northwest and making sure a surprise attack, like the one that happened at Pearl Harbor, would not happen again. (Courtesy of Air Force Historical Research Agency.)

The 25th Air Division had its own aircraft control and warning system, equipped with radar and jet fighters. Airmen used status boards and plotting tables in early 1952 to locate aircraft flight paths, allowing them to quickly tell the headings, altitude, and number of each aircraft being tracked. (Courtesy of Air Force Historical Research Agency.)

This photograph, taken on August 28, 1952, shows the interior of the new "Alert Hangar." Large overhead doors are visible on both sides of the photograph. A few tools lean against the far wall, built with corrugated metal sheeting. The Alert Hangars were used by the 325th Fighter-Interceptor Wing, which at the time was flying the F-94 Starfire. (Courtesy of Tacoma Public Library.)

The Alert Hangar at McChord is seen here on August 28, 1952. Four large overhead doors are stretched across the front of the building. The structure was built out of steel, bolted to a reinforced concrete pad, and covered in deeply corrugated siding. The alert hangars had central crew quarters, with opening front and rear doors for each aircraft pocket. (Courtesy of Tacoma Public Library.)

This photograph, taken on December 8, 1953, shows four members of the 465th Fighter-Interceptor Squadron, 325th Fighter Wing, 25th Air Division at McChord AFB in front of a North American F-86D Sabre "Dog." The 325th Fighter Wing had converted to the F-86D from the Lockheed F-94 Starfire. The 465th Fighter-Inspector Squadron (FIS) was a new squadron that was stood up earlier that year. In 1953, the F-86D Sabre was the fastest regulation military aircraft in the world, with a recorded speed of 715.697 mph. The pilots shown here are, from left to right, Lts. Thomas E. Randell, Franklin B. Brigham, Sam Huser, and Edwin Liddell. (Courtesy of Tacoma Public Library.)

McChord Air Force Base underwent another large-scale construction period in the 1950s. Buildings erected during World War II were either replaced or upgraded. The new Military Air Transport Service Air Freight building is seen here in 1952. (Courtesy of Air Force Historical Research Agency.)

This is another photograph of the new MATS Air Freight building in 1952. During World War II, the need to move cargo and people quickly became clear, and the Korean conflict reinforced that mission. Since large amounts of cargo needed to be moved, warehouses needed to be built to keep the cargo dry and secure until it could be moved to its final destination. (Courtesy of Air Force Historical Research Agency.)

Part of the upgrade to McChord AFB took place in Hangar No. 2 in February 1953. New lighting, visible in the background, was installed in the ceilings and walls. This photograph illustrates the enormous size of the hangar, which here accommodates a C-124 and a C-54. The size of these aircraft can be determined in comparison to the four men seen at center. (Courtesy of Tacoma Public Library.)

Shown here in 1955 is the A-frame used for removing and replacing the nose ramps on the C-124C aircraft. The C-124, or "Old Shakey," featured loading doors and hydraulic ramps in the nose that gave it a "clamshell" appearance when opened. The C-124 measured 48.3 feet tall and could carry a maximum of 66,348 pounds, or 200 troops or 127 litters (stretchers used to transport sick or wounded patients). The C-124s were assigned to McChord Air Force Base from 1951 to 1969. (Courtesy of Air Force Historical Agency.)

Here are interior (above) and exterior (below) views of the newly modified Douglas C-118A Liftmaster. The civilian equivalent of the DC-6, it was operated by the 33rd Air Transport Squadron (ATS), 1705th Air Transport Group, in 1955. A summary of the 1705th Air Transport Group mentions that the 33rd ATS had completed 333,000 flight hours in 12 C-54s in 34 months without an accident. The squadron also went on to fly six round-trips from Travis AFB, California, to Hickam AFB, Hawaii, in five and a half days. The C-118 was at McChord AFB from 1955 to 1960. (Above, courtesy of Air Force Historical Research Agency; below, courtesy of McChord Museum.)

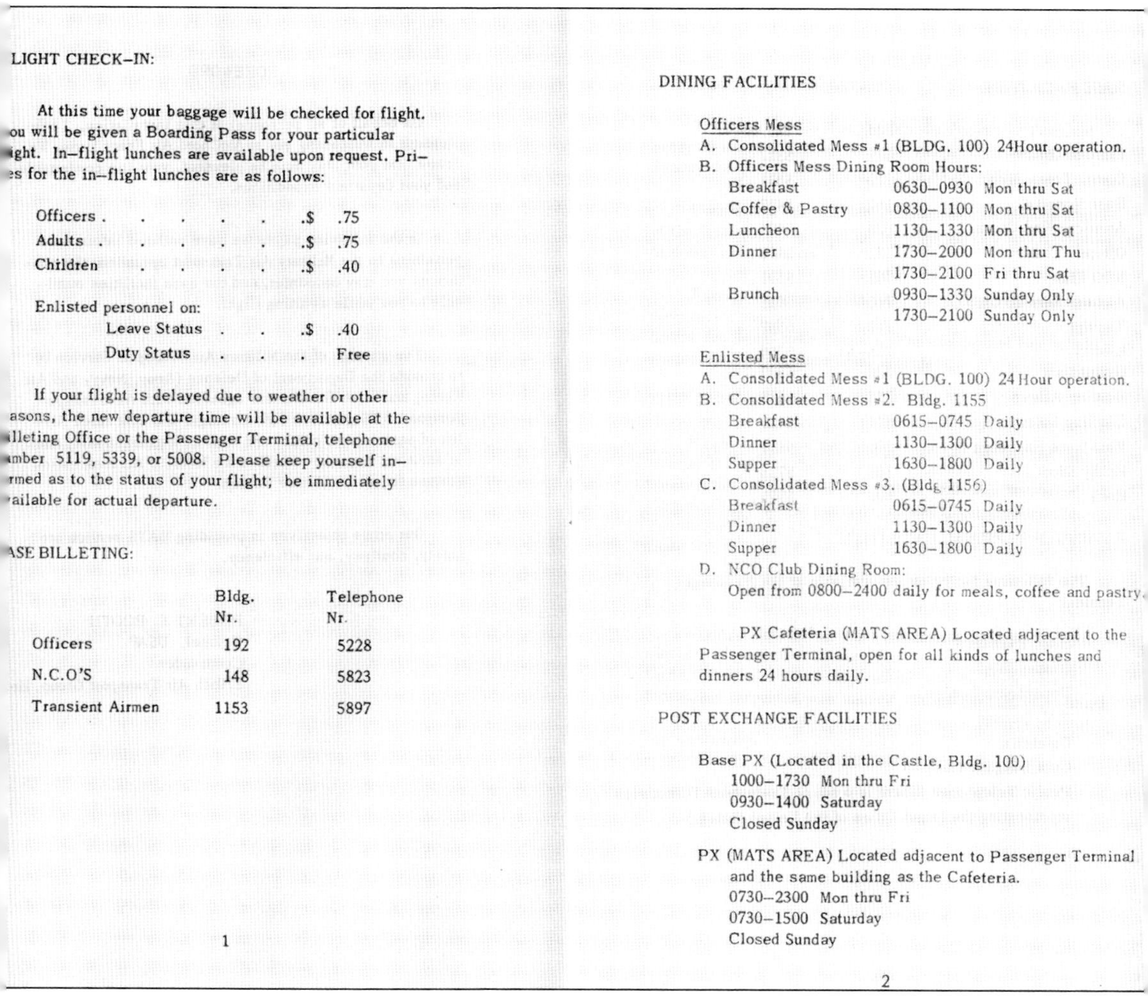

LIGHT CHECK–IN:

At this time your baggage will be checked for flight. ou will be given a Boarding Pass for your particular ight. In–flight lunches are available upon request. Pri–es for the in–flight lunches are as follows:

Officers	.$	.75
Adults	.$	.75
Children	.$	.40

Enlisted personnel on:

Leave Status . . .	.$	.40
Duty Status		Free

If your flight is delayed due to weather or other asons, the new departure time will be available at the illeting Office or the Passenger Terminal, telephone umber 5119, 5339, or 5008. Please keep yourself in–rmed as to the status of your flight; be immediately ailable for actual departure.

ASE BILLETING:

	Bldg. Nr.	Telephone Nr.
Officers	192	5228
N.C.O'S	148	5823
Transient Airmen	1153	5897

1

DINING FACILITIES

Officers Mess

A. Consolidated Mess #1 (BLDG. 100) 24Hour operation.
B. Officers Mess Dining Room Hours:

Breakfast	0630–0930	Mon thru Sat
Coffee & Pastry	0830–1100	Mon thru Sat
Luncheon	1130–1330	Mon thru Sat
Dinner	1730–2000	Mon thru Thu
	1730–2100	Fri thru Sat
Brunch	0930–1330	Sunday Only
	1730–2100	Sunday Only

Enlisted Mess

A. Consolidated Mess #1 (BLDG. 100) 24 Hour operation.
B. Consolidated Mess #2. Bldg. 1155

Breakfast	0615–0745	Daily
Dinner	1130–1300	Daily
Supper	1630–1800	Daily

C. Consolidated Mess #3. (Bldg. 1156)

Breakfast	0615–0745	Daily
Dinner	1130–1300	Daily
Supper	1630–1800	Daily

D. NCO Club Dining Room:
Open from 0800–2400 daily for meals, coffee and pastry.

PX Cafeteria (MATS AREA) Located adjacent to the Passenger Terminal, open for all kinds of lunches and dinners 24 hours daily.

POST EXCHANGE FACILITIES

Base PX (Located in the Castle, Bldg. 100)
1000–1730 Mon thru Fri
0930–1400 Saturday
Closed Sunday

PX (MATS AREA) Located adjacent to Passenger Terminal and the same building as the Cafeteria.
0730–2300 Mon thru Fri
0730–1500 Saturday
Closed Sunday

2

Anyone who has flown Space Available, Space A, or taken a "HOP" will appreciate some of the information printed in this 1956 Passenger Departure Information brochure of the 1705th Air Terminal Squadron. On another page in the brochure, women passengers are advised that wearing slacks is permissible for comfort. It also tells passengers that their baggage is limited to 65 pounds unless authorized otherwise on their orders, and that they should not place their travel documents or personal records in their stowed baggage. One last piece of advice was to observe the "Fasten Seat Belt" and "NO SMOKING" signs in the rear of the aircraft. Similar advice is given to those today flying with Air Mobility Command. (Courtesy of Air Force Historical Research Agency.)

U.S. AIR FORCE FU-045

Members of the 465th Fighter-Inspector Squadron spell out their new designation number, 318, in front of their F-86D Sabre "Dogs." On August 18, 1955, the 456th FIS was redesignated as the 318th FIS under Project Arrow. This operation was an Air Defense Command special project designed to return wartime honors to the command. Under this reactivation of historically significant units, the 325th Fighter Wing changed to a group. Earlier, on February 6, 1952, the 325th Fighter Wing was deactivated and replaced with the 4704th Air Defense Wing, absorbing all personnel and equipment of the 325th. More than a year later, on February 16, 1953, the 4704th was redesignated the 567th Air Defense Group. This group, in turn, was deactivated on August 18, 1955, and replaced by the 325th Fighter Group. (Courtesy of McChord Museum.)

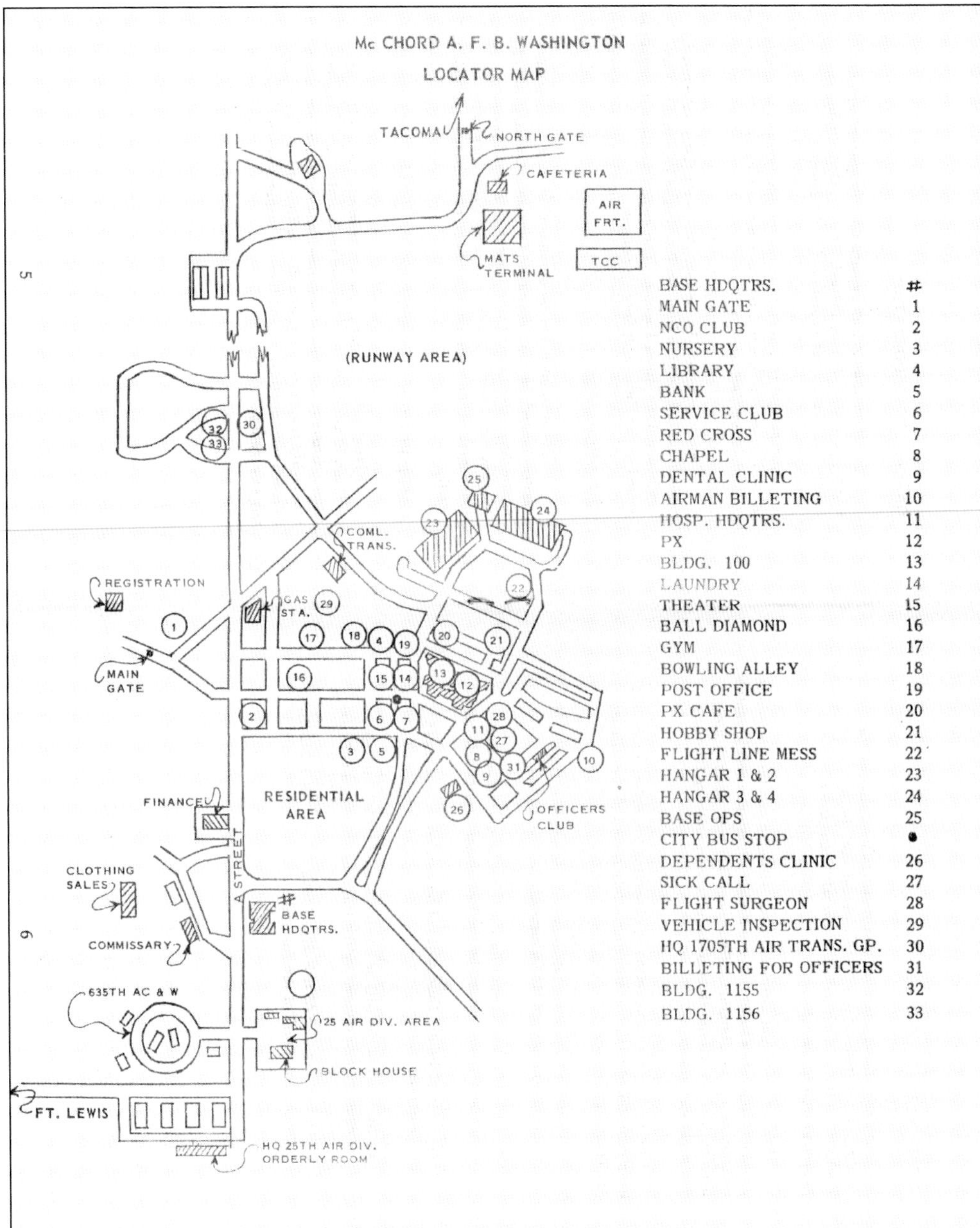

Troops taking a "HOP" to a new base often wondered where the base exchange was located, or where to grab a quick bite to eat while waiting for their next flight. The 1705th Air Terminal Squadron included this map in the Passenger Departure Information brochure in 1956 to assist passengers in finding their destination while either passing through McChord AFB or PCSing (Permanent Change of Station) to the area. This map also serves as a time capsule, showing how McChord AFB was laid out during the 1950s. (Courtesy of Air Force Historical Research Agency.)

This 1956 aerial photograph of the Air Terminal Squadron area offers a glimpse of the original, pre-1938 Tacoma Field hangar (foreground). The McChord AFB north gate can be seen center. Note the C-124 parked on the center right, along with a C-118 (lower right) and a PBY (lower left) at the bottom of the photograph. (Courtesy of Air Force Historical Research Agency.)

Passenger Service personnel wore new blue and white coveralls. This late-1950s photograph of the Passenger Terminal interior documents the early initiatives to identify Passenger Service personnel to military travelers with distinctive uniforms. It also illustrates Air Force efforts to distinguish the newly created service as a separate entity from the Army with different uniforms. (Courtesy of Air Force Historical Research Agency.)

Even with the rectangular addition, the original World War II–era modular building still retains its iconic image. These "temporary" buildings designed for their ease and speed of construction still dot US military installations today and served a wide variety of purposes before World War II, as evident in this photograph of the Passenger Terminal McChord Air Force Base in the late 1950s. (Courtesy of Air Force Historical Research Agency.)

This image of the loading dock illustrates a bygone era in military logistics where cargo was moved in small quantities and required much manual labor. Today's military freight terminals are complex facilities with sophisticated handling systems moving large quantities of cargo quickly. Note the elusive Mount Rainier in the background. (Courtesy of Air Force Historical Research Agency.)

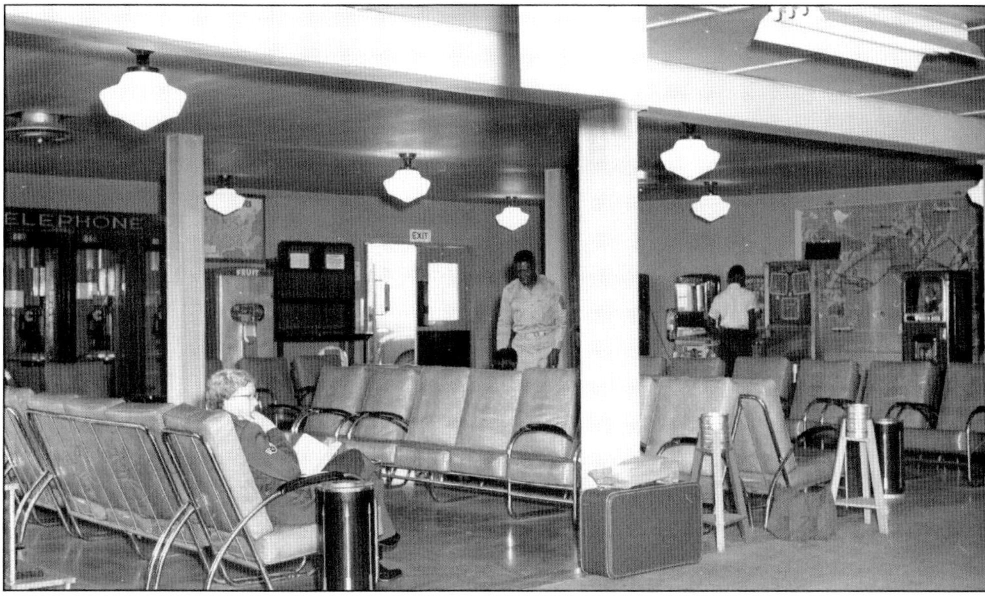

The McChord Passenger Lounge before remodeling in the 1950s resembled present-day passenger lounges. A notable difference is the abundant presence of ashtrays in the foreground that have all but disappeared except for a few designated areas in modern airports. Also, note the requirement for military personnel on leave to travel in uniform, another requirement that has faded into history. Along with the option to purchase insurance, as seen in the photograph below that was taken before a flight. (Both courtesy Air Force Historical Research

Figure 7 - The Passenger Service Counter before remodeling.
At the left is the new Flight Insurance Counter.

According to the 318th Fighter-Interceptor Squadron history, *Defenders of the Great Northwest*, its mission was "to intercept, identify and destroy enemy aircraft and airborne missiles penetrating the assigned area of responsibility." The two airmen in this c. 1957 photograph hold the wing of the newest Air Force jet, the Convair F-102A Delta Dagger, which replaced the North American F-86 Sabre. Note that the back of the airmen's flight jackets and their helmets display the same tail flash emblem as appears on the jet. This "Mach Wave" design was created by T.Sgt. Bruce McElhaney in 1957 to honor the plane's speed and future capabilities. The 317th FIS received the F-102, McChord AFB's first "supersonic" jet, on December 13, 1956, a year before the 318th FIS. Both squadrons, along with the 64th FIS, defended the Pacific Northwest during various years of the Cold War. (Courtesy of Tacoma Public Library.)

In late 1957 and early 1958, new family housing had been built, and the units were ready for occupancy shortly after January 1, 1958. There were 110 units built, 70 houses for enlisted members (above) and 40 for officers. The new housing area was located near the base's western boundary, on Woodbrook Road adjacent to Highway 99. Each unit cost around $13,000, which included utilities and landscaping; the money came from the 1956 Military Construction program appropriations. In 1958, the first on-base school, to accommodate 500 students, was under construction, with a scheduled completion date of September 1959. The below photograph offers a broad view of the new housing area. (Courtesy of Air Force Historical Research Agency.)

According to the First Quarter (FY-60) Historical Report, the contract to update Project Number McC 149-9, the dining hall in building 1155, was awarded for $14,257.00. The update to the facility began on July 6, 1959, and was completed on September 16, 1959. This was another example of base modernizing that took place during the 1950s. (Courtesy of Air Force Historical Research Agency.)

The NCO Club, or Noncommissioned Officers' Club, is seen here in the 1950s. The McChord Air Force Base Directory and Guide for 1958 states that membership to this club is restricted to those holding the rank of staff sergeant and above. The club offered a complete bar, lounge, television room, bingo on Thursday nights, dancing on the weekends, and special events during the holidays. (Courtesy of Air Force Historical Research Agency.)

The Women Officers Quarters is seen here in 1953. When it came to women's status and involvement, the 1950s was a different time at McChord AFB, in the United States military, and in American society in general. In 1951, the military preferred single women, women were automatically discharged for becoming pregnant, and mothers of children under the age of 18 were not allowed to volunteer for service. (Courtesy of Air Force Historical Research Agency.)

In 1958, the base theater was located in building 742. There were two performances nightly, one at 6:00 and one at 8:00. A matinee was offered at 2:15 p.m. on Saturday and Sunday. As is the case today, patrons had to show their military identification card to gain admittance to the theater. (Courtesy of Air Force Historical Research Agency.)

The Mobile Exchange, operated by the Post Exchange (PX) on McChord Air Force Base, is seen here in 1957. It resembles a modern-day Class Six Shoppette, where members can purchase various beverages on base. (Courtesy of Army and Air Force Exchange Service.)

A clerk works in the post exchange, which was located at the northeast end in the basement of building 100, otherwise known as the "Castle." As seen in this 1958 photograph, the PX was adequately stocked with various goods and, according to the 1958 *Base Guide*, was open from 10:30 a.m. to 5:30 p.m. Monday through Friday, on Saturday from 9:30 a.m. to 2:00 p.m., and was closed on Sunday. (Courtesy of Army and Air Force Exchange Service.)

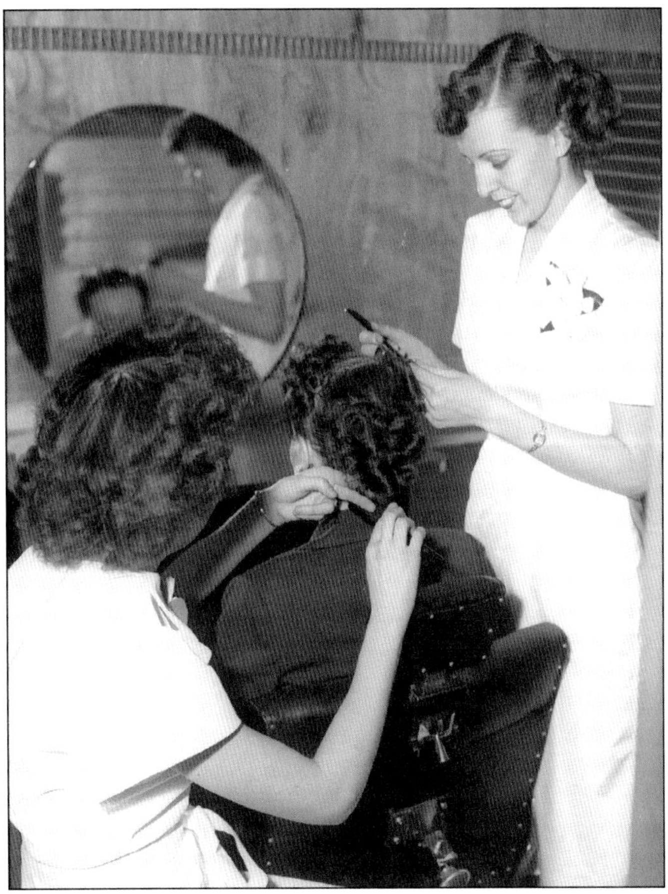

This 1958 photograph is probably of the main PX café, located behind the Castle, in building 108. The café was open from 6:30 a.m. to 11:30 p.m. and offered coffee, sandwiches, and other dinner menu items. Another eatery on base was the MATS cafeteria, which was open 24 hours a day. (Courtesy of Army and Air Force Exchange Service.)

A woman is having her hair worked on by two stylists at the base's beauty shop in 1958. The barbershop was located in the Castle. The beauty shop was probably also located in the Castle. (Courtesy of Army and Air Force Exchange Service.)

Workers on scaffolding look down at crew members below on August 27, 1957, during the construction of the Semi-Automatic Ground Environment (SAGE) building. The system was labeled Semi-Automatic because the commander would utilize information from the computer, which was built by IBM. (Courtesy of Tacoma Public Library.)

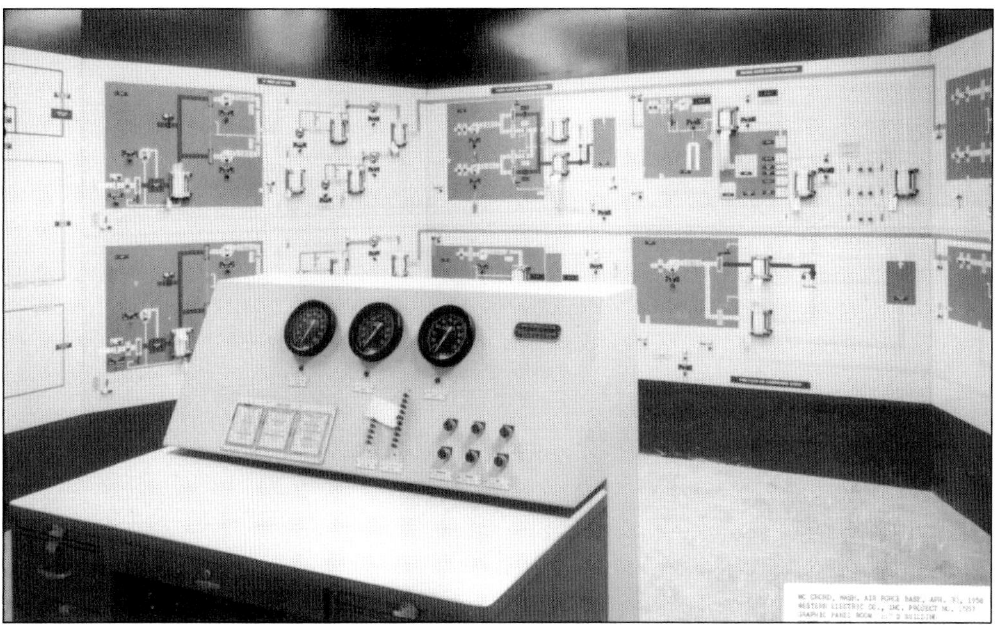

The Western Electric Co. of New York City had been working on several projects on base. Among them was the graphic panel room in 117 D Building in the SAGE building, shown here on April 30, 1958. The control board was carefully labeled, with some switches connected to the second-floor computer, others to the fresh air system and the fourth floor air-conditioning system. (Courtesy of Tacoma Public Library.)

This is the SAGE facility as it appeared in the 1958 *Base Guide*. The facility housed large computers that kept track of air traffic, displayed weather conditions, told the current status and location of defensive weapons, and directed and controlled jet fighters and missiles with the mission to defend the United States from aerial attack. A year earlier, Canada and the United States agreed that an air attack against the North American continent would be opposed by both countries. The two nations formed the North American Air Defense Command (NORAD). The SAGE headquarters at McChord were maintained by a joint US and Canadian force and defended a total air space of 822,000 miles. (Courtesy of McChord Museum.)

Four

Maintaining Global Peace

"I christen thee." Marsha McLean, Clover Park's Daffodil Princess, spoke these words on September 9, 1960, as she broke a bottle of champagne on the *City of Tacoma*, the first Boeing KC-135 Stratotanker to arrive at McChord AFB. This ushered in a new mission, Strategic Air Command (SAC). Standing next to McLean is Deputy Mayor James S. Porter (center) and Col. Kenneth C. Diehl, commander of the 22nd Air Refueling Squadron, 92nd Bombard Wing (Heavy). (Courtesy of McChord Museum.)

Irish troops disembark from a 62nd Troop Carrier Wing, MATS, C-124 Globemaster in the summer of 1960. The Irish troops were part of a UN contingent participating in an airlift that was originally called Operation Safari and then Operation New Tape. This undertaking was the largest airlift since the Berlin blockade and is considered to be one of the US Air Force's greatest peacetime accomplishments. Within 25 minutes after H-hour, on July 19, 1960, five aircraft of the wing departed McChord for the Republic of the Congo. In total, 18 C-124s took part in the airlift to evacuate personnel and bring back UN troops and equipment. In early September 1960, the 62nd's participation in Operation New Tape ended. (Courtesy of McChord Museum.)

Members of Detachment 5 at McChord Air Force Base pose with a Kaman HH-43 Huskie rescue helicopter. The gathering marked a significant accomplishment: 2,000 hours flown without the loss of a member or an aircraft. The detachment was stationed at McChord from 1961 to 1966 under the Western Air Rescue Center and in 1966–1969 under the Western Aerospace Rescue and Recovery Center, with its headquarters at Hamilton AFB, California. The detachment assisted in numerous recoveries in the area. The helicopter's ability to reach a service ceiling of 25,000 feet, carry a crew of four, including two rescue crew members, and stay in the air up to three hours, allowed the detachment to recover many stranded people. (Courtesy of McChord Museum.)

According to an article in the *Tacoma News Tribune*, in July 1953 the commissary burned to the ground due to a lack of water. The loss was estimated as $124,000. The 1958 *Base Guide* states that the commissary had been recently erected and was located in building 505. By 1962, the *Base Guide* stated that the commissary was located in building 577. In both guides, the commissary is listed as being open Tuesday thru Friday from 10:00 a.m. to 5:00 p.m. (5:30 p.m. in 1962); Saturday from 10:00 a.m. to 3:00 p.m.; and closed Sunday and Monday. Now, however, the commissary is open seven days a week and has later closing times. These photographs were taken on July 5, 1961. Above, a ribbon-cutting ceremony celebrates the opening of the commissary at its new location in building 577. Compared to the old facility, the new one offered customers better service and more space. (Courtesy of Air Force Historical Research Agency.)

This iconic image shows three Century-Series interceptors, a popular name for fighter model F-100–F-106 aircraft, assigned to the 25th Air Division. The planes are flying past Mount Rainer around 1964–1965. The aircraft are, from left to right, Convair F-102A 56-1333 from the 64th FIS, Paine Air Force Base, Washington; McDonnell CF-101B, Tail No. 17460 from the 409th Squadron, RCAF Station Comox, BC; and a Convair F-106A 59-0052 from the 318th Fighter Inceptor Squadron. Paine AFB, located near Everett, Washington, became home to the 64th FIS, which had moved from McChord on February 8, 1960. The 409th Squadron, RCAF Station Comox, BC, was comprised of Royal Canadian Air Force aircraft that provided air defense for Canada's west as part of NORAD. (Courtesy of McChord Museum.)

Miss Washington 1966, Sandra Marth, christens a Lockheed C-141A 65-000277, the Tacoma Starlifter, at an official acceptance ceremony on August 9, 1966. A new era had begun at McChord AFB with the arrival of the C-141 as part of the 62nd Military Airlift Wing (MAW), 8th Military Airlift Squadron. The aircraft would remain at McChord until 2002. The C-141, "Old Droopy," which replaced the aging C-124, "Old Shakey," received its nickname because the 160-foot wingspan drooped low while on the ground, rising only when in the air. In the late 1960s, the C-141 could haul up to 70,000 pounds of cargo, had a cruising speed of 566 miles per hour, and had an incredible range, especially with in-flight refueling. During the Vietnam War era, the C-141 was used to transport cargo and troops to the Far East and Alaska. (Courtesy of McChord Museum.)

On April 2, 1968, Brig. Gen. John A. Rouse presented the Purple Heart to nine-year-old Frederick R. Sebers, on behalf of his father, T.Sgt. Frederick T. Sebers, who was killed in the Republic of Vietnam. Sergeant Sebers, who was born in Everett, Washington, died on November 7, 1967, at Madigan Army Hospital in Tacoma, Washington, from wounds he had received on September 8, 1967. A Claymore mine detonated near Sebers's dining table at the NCO club in Nha Trang, Vietnam, causing life-threatening wounds. Jane S. Sebers and her daughter, Sylene, 12, look on during the presentation. (Courtesy of Air Force Historical Research Agency.)

Continuing its role in Vietnam and at the center stage of world events, the C-141 and McChord AFB's 62 MAW played an important role into the early 1970s. 62nd MAW flew to Hanoi, North Vietnam, in support of Operation Homecoming, the return of POWs from Vietnam from February 12 to April 1, 1973, using C-141s. In this photograph, US Navy Cdr. James Glenn Pirie speaks to the crowd on the side of a C-141 65-0258 at Clark AFB on February 20, 1973. Red Cross 62nd Military Airlift Wing and 939th Military Airlift Group insignias are displayed on the tail of 0258. All aircraft directly involved in the POW mission wore Red Crosses on their tails. (Courtesy of McChord Museum.)

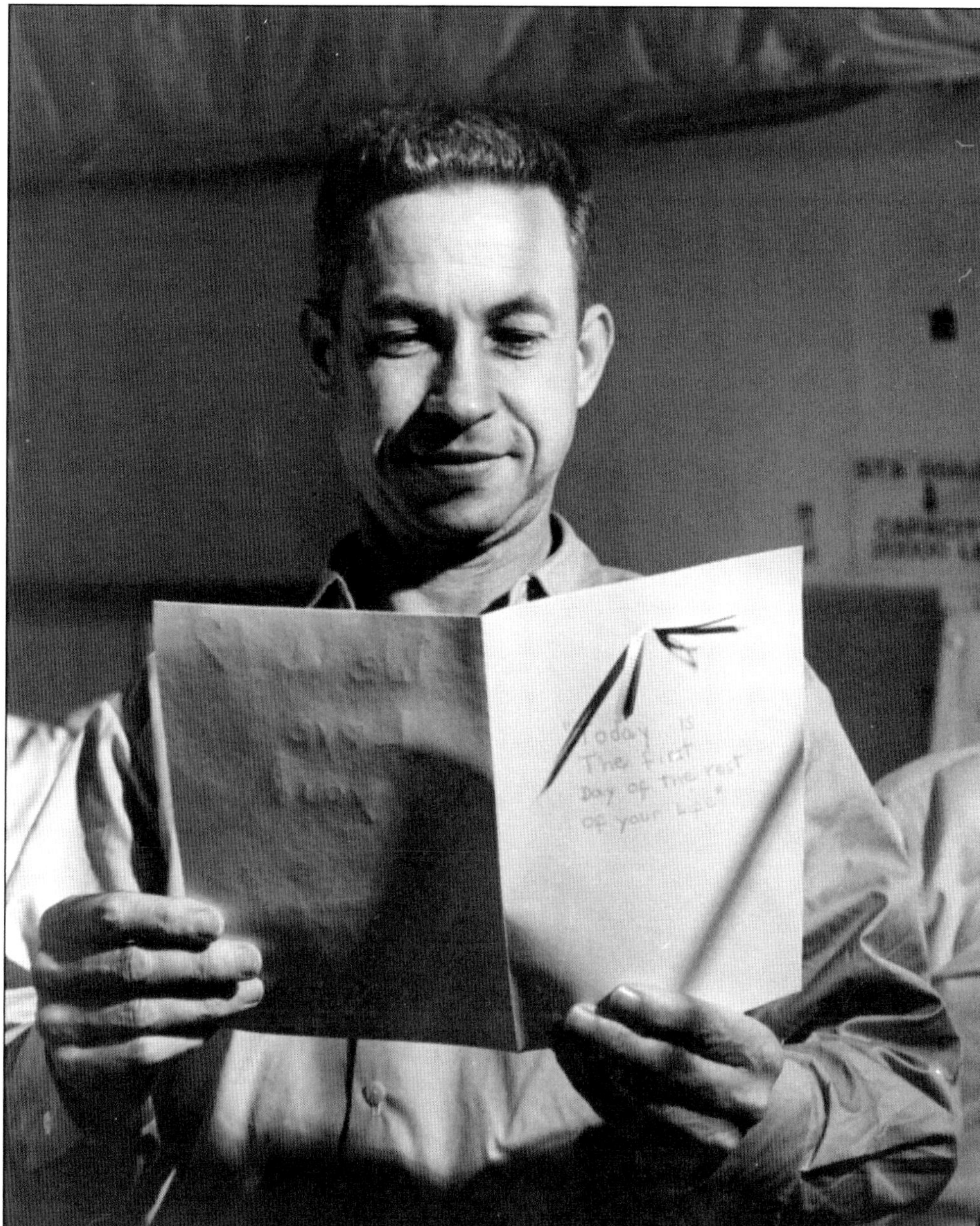

One of the prisoners that returned from North Vietnam on a 62nd MAW C-141 on March 4, 1973, was Col. Norm Gaddis, the senior ranking POW. The card that Colonel Gaddis is holding reads, "Today is the first day of the rest of your life." Gaddis was shot down on May 12, 1967, and was not released until March 4, 1973. He was later awarded the Air Force Distinguished Service Medal, one of several medals awarded to him for his sacrifice during the Vietnam War. His AF Distinguished Service Medal citation reads, "General Gaddis displayed professional competence, unwavering devotion and loyalty to his country in the execution of his duties in staff and command positions while in potentially volatile daily contact with the Vietnamese guards and officers." He was promoted to brigadier general on June 1, 1974, and retired on June 1, 1976. (Courtesy of McChord Museum.)

Another POW involved with Operation Homecoming was Cdr. Edward M. Martin. He was brought home on March 4, 1973, by the 62nd MAW. Commander Martin was born in Savannah, Georgia, on September 30, 1931, and graduated from the Naval Academy in 1954. On July 9, 1967, Martin was leading a flight of A4 Skyhawks from the carrier USS *Intrepid* when he was shot down. He ejected safely, but was quickly captured. Martin spent five years and eight months in captivity in the immediate Hanoi area. He retired from the US Navy as a vice admiral in 1989. (Courtesy of McChord Museum.)

An F-106 Delta Dart is parked on the ramp at McChord Air Force Base in the mid-1970s. According to the 318th Fighter-Interceptor Squadron website, the 318th wanted to "freshen up the unit's tails flash" with the new marking of a "Starburst," as seen in this photograph. The starburst resembles the earlier 1960s F-106 tail flash, but in this case, eight rays extend from the center disk and reach the edges of the tail. (Courtesy of McChord Museum.)

A F-106 Delta Dart flies past Mount Rainer. Note the "Spirit of 76" icon painted on the tail. This special paint scheme, in celebration of the nation's bicentennial, was authorized by the US Air Force for one plane from each squadron. The 318th FIS chose "Freedom Bird" No. 58-0776, dedicating it during the squadron's 1975 Hughes Trophy ceremony on January 24, 1976. (Courtesy of McChord Museum.)

Aircraft No. 58-0776 (also pictured in flight at the top of this page) is shown here in 1976 with the bicentennial tail flash and "The Freedom Bird" painted on the external fuel tanks. These four men represented the 318th FIS at the annual Weapons Loading Competition at Tyndall Air Force Base, Florida. They are, from left to right, S.Sgt. Ted Standifer, Sgt. Carolos Martinez, Sgt. Richard Keyes, and Sgt. Michael Budzisz. (Courtesy of McChord Museum.)

In 1975, the 62nd MAW took on a new mission and a new aircraft, the Lockheed C-130 Hercules, which remained at McChord AFB until 1989. The C-130s were assigned to the 36th Tactical Airlift Squadron (TAS) that had come from Langley Air Force Base, Virginia. The 62nd MAW was the only Air Force wing to fly both the C-141 (pictured in the background, used for strategic airlift missions) and the C-130 (pictured in the foreground, used for tactical airlift missions). (Courtesy of McChord Museum.)

In May 1980, Mount St. Helens, which had been dormant for 123 years, erupted. A C-130 crew from the 36th TAS provided communications support during the search for survivors. One week later, the mountain erupted again. With the threat of ash drifting toward McChord AFB, all of the base's flyable aircraft were evacuated, in four hours and eighteen minutes. The volcano erupted again in August, leaving a noticeable coat of ash on McChord AFB. (Courtesy of McChord Museum.)

The Burger Barn, or Hamburger Barn, seen here in 1979, was operated by the Army Air Force Exchange Service (AAFES). Anyone who was stationed at or lived close to McChord Air Force Base remembers the delicious hamburgers that were cooked in this little red building. In the late 1990s, AAFES no longer operated the Burger Barn, replacing it with a commercial franchise establishment that still operates today. (Courtesy of Army Air Force Exchange Service.)

An unidentified AAFES worker places sandwiches out for the day in 1971. (Courtesy of Army Air Force Exchange Service.)

A four-ship formation of F-106As from the 318th FIS fly close to Mount Rainer in 1982. Note, underneath the belly of the two near planes, a "six shooter" gun package. This is a six-barrel cannon located in the internal weapons bay. (Courtesy of Ernest White.)

A F-106A from the 318th Fighter-Interceptor Squadron flies by Mount Rainer sometime in 1980–1981. This particular fighter, No. 59-0147, was the second-to-last one delivered to the US Air Force, in November 1960. It was also the last one to leave McChord AFB, as the 318th FIS converted to the McDonnell Douglas F-15 Eagle. (Courtesy of Ernest White.)

GARY

This photograph, taken of a pilot in the backseat of a F-106A two-seat trainer, shows two F-106As in formation. The peak of Mount Rainer is in the background. The aircraft were returning from a local training mission in the early 1980s. (Courtesy of Ernest White.)

Four of the last five T-33 "T-Birds" fly in formation over Mount Rainer on February 3, 1988. Brig. Gen. John M. Davey, 25th Air Division commander, was piloting the lead aircraft. The second plane, flying slightly in front of the others, wears the general's star on the lower part of the tail. This was the farewell flight and retirement of the T-33 Shooting Star from the US Air Force. (Courtesy of Ernest White.)

This is the airfield at Point Salines, Grenada, as it looked when a McChord AFB Starlifter, the first C-141, landed there during Operation Fury. This mission involved the 62nd Military Airlift Wing beginning at 12:30 a.m., Saturday, October 22, 1983. The mission involved evacuating American medical students on Grenada, whose safety became a concerned when a takeover by a pro-Marxist regime endangered their safety. In addition, the Cubans were building a major runway that posed a military threat on the island at Point Salines. As the mission developed, the 62nd MAW moved its C-141s to Pope Air Force Base, North Carolina, to pick up the Army's 82nd Airborne Division, on October 24. En route to Point Salines, the 82nd Airborne's commander was ordered to do a landing instead of an airdrop. The McChord crews carried out the order by landing on a runway about 8,500 feet in length. The C-141 was on the ground for only seven minutes with engines running while it offloaded the 82nd Airborne Rangers. (Photograph by Capt. Joseph Leister; courtesy of Air Force Historical Research Agency.)

The 62nd Security Police Squadron, led by Lt. Col. Thomas McKinsey, transferred prisoners from American to Cuban control on November 3–11, 1983, during Operation Urgent Fury. Here, security police stand in front of a Cubana Airlines craft. The Soviet-built Ilyushing IL-62M carried prisoners from Barbados to Cuba. (Courtesy of Air Force Historical Research Agency.)

Cuban authorities (right) and an International Red Cross representative (center) worked with an American official (seated at left) to transfer Cuban prisoners brought from Grenada to Barbados in the late fall or early winter of 1983. (Courtesy of Air Force Historical Research Agency.)

This beautiful photograph illustrates McChord's air heritage with three cargo aircraft based out of McChord AFB flying in front of Mt. Rainier. The plane on the left is a C-141, the one at center is a C-130, and the third plane is a C-124C, No. 52-0994. The photograph was taken on October 9, 1986, during a nonstop flight from Selfridge Air National Guard Base, Michigan, to McChord AFB. After several months of work by McChord Museum volunteers and members of the Air Force and the Air Force Reserve, aircraft No. 52-0994 was put on display on Heritage Hill at McChord Air Force Base. The C-141 replaced the C-124. The C-130 was based out of McChord AFB at the same time as the C-141. Furthermore, this was the last flight of any C-124 Globemaster II aircraft shown in this photograph. (Courtesy of McChord Museum.)

Snow blows around a 62nd MAW C-141 during Brim Frost, a Joint Chiefs of Staff exercise held every other winter in Alaska. This photograph was taken on January 30, 1989. The purpose of the exercise was twofold: first, it provided a scenario for those involved in hostile ground and air operations throughout the state of Alaska; and second, it provided valuable winter training for ground and air units of the 62nd Military Airlift Wing. (Photograph by Frederick A. Johnsen; courtesy of Air Force Historical Research Agency.)

A Canadian deicing truck moves toward a C-141B from McChord AFB during Brim Frost 1989, as wind and snow blow around the field. According to weather reports of the time, it was 50 degrees below zero Fahrenheit. Brim Frost was held from January 4 through February 7, 1989. (Photograph by Frederick A. Johnsen; courtesy of Air Force Historical Research Agency.)

Evacuees from Clark Air Base and Subic Bay Naval Base, Philippines, arrive at McChord AFB on June 18, 1991, following the eruption of Mount Pinatubo on June 9, 1991. Three chartered commercial planes were used to bring the evacuees to McChord AFB. The first plane, carrying 187 passengers, arrived at 4:40 p.m. on a Thursday; the second aircraft landed at 11:00 p.m.; and the third landed the following day, around 12:30 p.m. Six flights of C-141s were supposed to land at McChord AFB, but they were diverted to Travis AFB, California, to help ease the burden at McChord. The repatriation operation was called Fiery Vigil and lasted until July 3, 1991. (Courtesy of McChord Museum.)

Many of the evacuees arriving at McChord Air Force Base had left most of their possessions behind in the Philippines. While at McChord, evacuees received a variety of assistance, including airplane tickets to homes and duty stations around the United States. According to the *Tacoma News Tribune*, the Tacoma-Pierce chapter of the American Red Cross planned to use about 260 volunteers a day to escort people through the processing center and provide emergency services, such as food, refreshments, and childcare. McChord AFB was chosen for its location on the Pacific Rim and the availability of commercial flights at Seattle-Tacoma International Airport. (Courtesy of McChord Museum.)

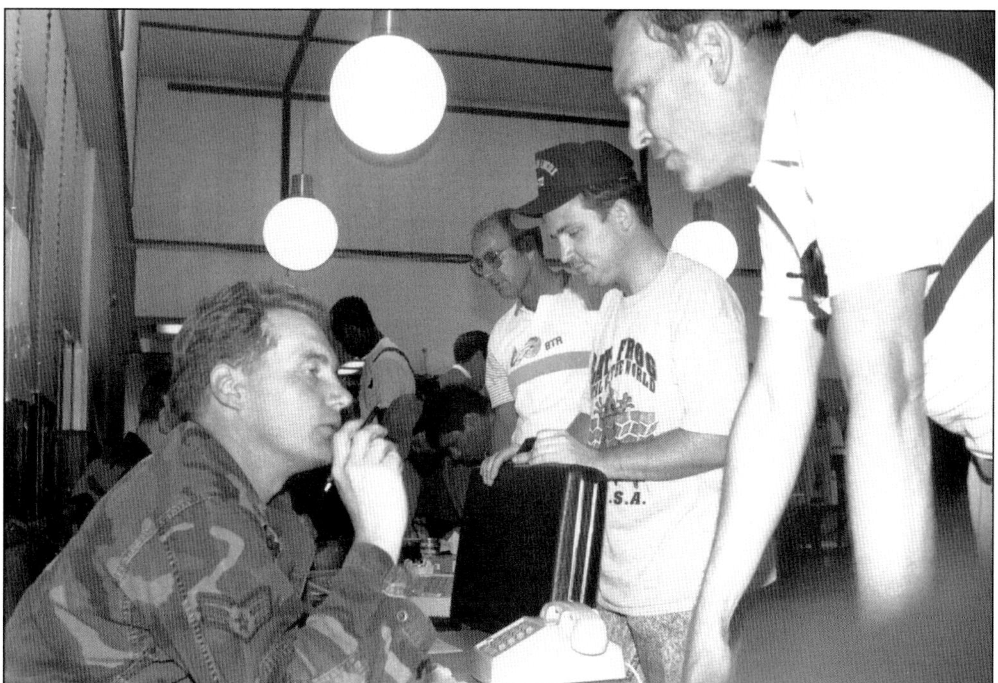

A Boeing C-17 Globemaster III is parked on the ramp during a Joint Base Lewis-McChord air show, July 21–22, 2012. Also visible is one of the original hangars, built before World War II. The C-17 arrived at McChord Air Force Base in 1999, replacing the previous cargo plane, the C-141. The C-17 can carry a maximum of 585,000 pounds, making it the current workhorse of the US Air Force. The C-17 has been involved in many global wartime and humanitarian missions. (Author's collection.)

On October 14, 2012, Airmen from the 62nd and 446th Airlift wings completed a milestone event flying the 500th mission to Antarctica in support of Operation Deep Freeze, the US military's support of the National Science Foundation. Almost 13 years to the date, the first McChord C-17 landed on the icy runway outside McMurdo Station, Antarctica, on October 15, 1999. (Courtesy of United States Air Force.)

Aircraft Stationed at McChord Field

Attack/Light Bomber	Years of Service
Douglas A-20 Havoc	1941
Douglas A-24 Banshee	1944–unknown
Fairchild Republic A-10 Thunderbolt II	1993–1994**
Lockheed A-29 Hudson	1942
Northrop A-17 Nomad	1940
Bomber	
Boeing B-17 Flying Fortress	1940–1943
Consolidated B-24 Liberator	1945
Consolidated LB-30 Liberator	1943
Douglas B-18 Bolo	1940–1941**
Douglas B-23 Dragon	1940–1941**
Martin B-26 Marauder	1941
North American B-25 Mitchell	1941–1942
Cargo/Transport	
Beech C-45 Expeditor	1943–1949
Boeing C-17 Globemaster III	1999–Present
Curtiss C-46 Commando	1952
Douglas C-47 Skytrain	1948–1961**
Douglas C-54 Skymaster	1950–1955
Douglas C-118 Liftmaster	1955–1960
Douglas C-124 Globemaster II	1950–1969**
Fairchild C-82 Packet	1947–1949**
Lockheed C-130 Hercules	1975–1989
Lockheed C-141 Starlifter	1966–2002**

Fighter/Pursuit	
Convair F-102 Delta Dagger	1956–1960**
Convair F-106 Delta Dart	1960–1983**
Lockheed P-38 Lightning	1942–1943
Lockheed F-94 Starfire	1950–1953
McDonnell Douglas F-15 Eagle	1983–1989**
North American F-51 Mustang	1951–1952
North American F-82 Twin Mustang	1947–1950
North American F-86 Sabre	1953–1957**
Northrop P-61 Black Widow	1947–1948
Republic P-43 Lancer	1941
Republic P-47 Thunderbolt	1944
Helicopter	
Bell H-13 Sioux	1949
Kaman H-43 Huskie	1961–1969, 1971–1972
Sikorsky H-19 Chickasaw	1953–1961
Liaison/Observation	
Fairchild Republic OA-10 Thunderbolt II	1993–1994**
Stinson L-5 Sentinel	1945
Reconnaissance	
Boeing WB-50D Superfortress	1960–1961
Search & Rescue	
Boeing SB-17 Flying Fortress	1947–1952?
Canadian Vickers SA-10A Catalina	1946–1950**
Grumman SA-16 Albatross	1953–1956
Tanker	
Boeing KC-135 Stratotanker	1960–1962
Trainer	
Beech AT-11 Kansas	1951
Convair T-29 Flying Classroom	1969–1975
Lockheed T-33 Shooting Star	1951–1988
North American (A)T-6 Texan	1946
North American T-39 Sabreliner	1970–1975
Stearman (Boeing) PT-17 Kaydet	1941–1942
Vultee BT-13 Valiant	1941

** Aircraft on display at McChord Museum Air Park

Bibliography

Anything, Anywhere, Anytime: An Illustrated History of the Military Airlift Command 1941–1991. Scott Air Force Base, IL: Headquarters Military Airlift Command, May 1991.

Bright, Charles D., ed. *Historical Dictionary of the US Air Force.* New York: Greenwood Press, 1992.

History of McChord Field 1938–1945. Maxwell Air Force Base, AL: 1974.

Hutcheson, Keith. *Air Mobility: The Evolution of Global Reach.* Vienna, VA: Point One and VII Publishing, 1999.

Johnsen, Frederick A. *Volume I, History of 62nd Military Airlift Wing McChord AFB, Washington October 1–December 31, 1983.*

Knaack, Marcelle Size. *Encyclopedia of US Air Force Aircraft and Missile Systems, Volume I: Post–World War II Fighters, 1945–1973.* Washington, DC: Office of Air Force History, 1978.

McCroskey, Lauren. *McChord Field Historic District.* Seattle: Center of Expertise, Preservation of Historic Buildings and Structures, Seattle District US Army Corps of Engineers, February 2008.

www.318thfis.com

www.af.mil/information/bios/

www.amcmuseum.org/history/airlifts/congo_airlift.php

www.mcchordairmuseum.org

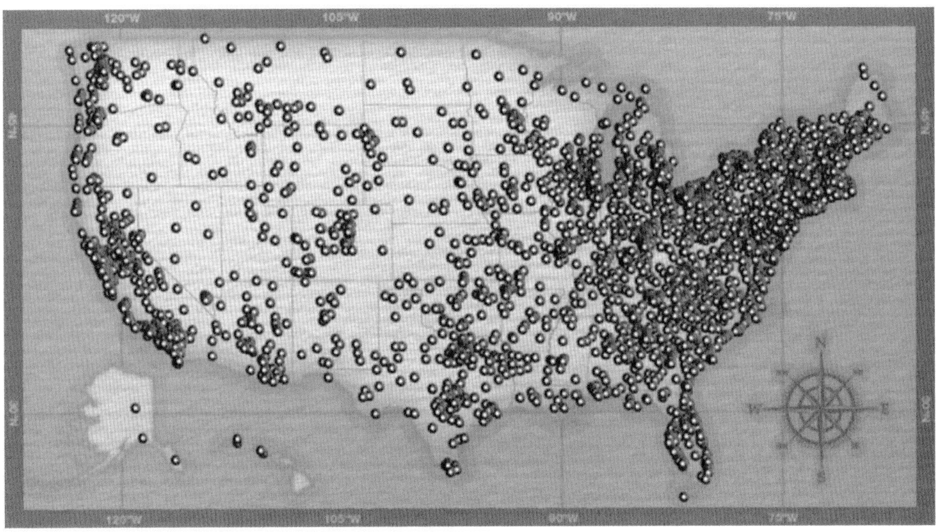